KTM Book $4.00

OUR LIVING WORLD OF NATURE

The Life of Rivers and Streams

Developed jointly with The World Book Encyclopedia

Produced with the cooperation of The United States Department of the Interior

OUR LIVING WORLD OF NATURE

The Life of Rivers and Streams

ROBERT L. USINGER

Published in cooperation with
The World Book Encyclopedia
McGraw-Hill Book Company
NEW YORK TORONTO LONDON

ROBERT L. USINGER *is Professor of Entomology and Chairman of the Division of Entomology at the University of California, Berkeley. He has taught a course in aquatic entomology for many years and is a specialist on several groups of aquatic insects. For a number of years he has conducted field work at the University of California Sagehen Creek Experiment Station, which he helped establish for the study of natural reproduction in trout. In addition to editing the large book* Aquatic Insects of California, *published by the University of California Press, he has written several other textbooks and monographs. He has traveled extensively in Africa, South America, and Southeast Asia and has worked at fresh-water biological stations in England, Finland, and Brazil, and on Lake Tanganyika in Africa.*

Library of Congress Catalog Card Number: 67–14852

1234567890 NR 721069876

66690

OUR LIVING WORLD OF NATURE

Contents

WATER ON THE MOVE 9

Life in the rapids 12; Many worlds in one 18; Sampling stream life 20; Torrents and waterfalls 24; Life along the bank 29; Life on the surface film 36; Breathing underwater 38; The beetle and the bubble 42; Insect gills 44; The miracle of metamorphosis 48; Mayflies molt twice 51; Life in the flowing water 57

FROM SOURCE TO SEA 59

Life in a spring 61; Hot springs and geysers 62; Life in hot springs 63; The water cycle 68; Rivers of darkness 73; Snowpack and surface water 74; Rivers of ice 76; River patterns on the land 81; Downstream 82; Otter antics 84; Of muskrats and mussels 88; Mussel habits 90; How mussels multiply 91; Life on the muddy bottom 92; The flight of the dragonflies 94; Millions of mayflies 96; A swarm of midges 97; River turtles 98; Fishes everywhere 102; The end of a journey 105

RIVERS OF LIFE 109

The river factory 110; Green plants everywhere 112; The green film of life 115; Along the assembly line 115; Food that floats 117; Food that falls in 119; Measuring productivity 119; Primary productivity in Silver Springs 121; The production pyramid 124; To catch a trout 127; More fish for fishermen 128; So many fish—and no more 130; Rivers and people 133; America's sick rivers 134; Kinds of pollution 136; The road to recovery 138; Too much too soon 140; New problems for our rivers 141; Can the problems be solved? 143; Rivers for people 145

LAND OF MANY RIVERS 151

The wild Hudson 155; The St. Lawrence 158; Lampreys in the Great Lakes 160; The father of waters 162; The Mississippi's course 164; Floods on the Mississippi 166; A better life in the Tennessee Valley 167; The Rio Grande 170; Across the Rockies 176; The Colorado River 177; North to the Columbia 184; River of fish 191; The beckoning rivers 199

APPENDIX

Rivers and Streams in the National Park System 203; How to Study Fresh-water Life 211; A Guide to Aquatic Insects 215; Vanishing Fishes 220

Glossary 222
Bibliography 227
Illustration Credits and Acknowledgments 228
Index 229

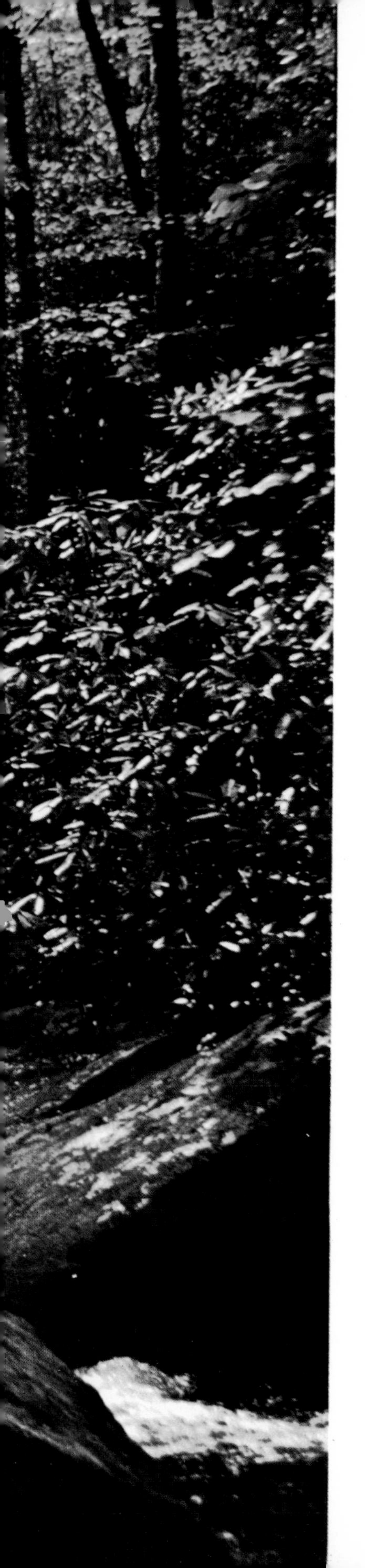

Water on the Move

Running water! As you hike through the cool, dim forest, you hear the sound of flowing water even before you see the stream. But as you push through a thicket, you find yourself suddenly at the edge of a ravine. There at the bottom a stream flows clear and cool, with a sound like rustling leaves. A foaming white cascade plunges across a rocky ledge, then loses itself in the rippled surface of a deep, dark pool. Bubbling between moss-covered rocks, swirling into placid backwaters, or gliding velvet-smooth across more level stretches, the stream rushes down its channel and disappears from sight around a bend.

As you scramble down the steep slope of the ravine, the whispering sound of the stream grows louder and the air becomes noticeably cooler. A startled deer leaps from a bed of ferns beside the stream and bounds easily up the opposite slope. A kingfisher, alarmed by the commotion, lets out a series of harsh rattling calls as it flies upstream to perch on a branch overhanging the water. If you were to wait long enough, you might see the ungainly blue and white bird swoop suddenly from its perch, dive into the water, and come up with a minnow or some other small fish squirming in its sturdy beak.

With the speed of a dive bomber, a belted kingfisher plummets from its perch and plunges into the shallows in a placid stream *(left)*. An instant later, it emerges with a fish clamped firmly in its beak *(right)* and returns to its perch, usually a dead branch overhanging the water. With a few sharp blows from its heavy beak, the bird will stun its squirming victim, swallow the fish head first, and then scan the water for telltale flashes that reveal the movements of other fish in the stream. Although trout fishermen sometimes resent these feathered competitors, conservation studies have shown that kingfishers usually have little effect on the populations of game fish. In return for the few desirable fish they capture, the big blue and white birds add color and excitement to the scene along streams and rivers.

But right now you are more interested in a brilliant blue-green dragonfly that darts past, patrolling a regular route along the stream. With its spindly legs folded to form a basket beneath its body, this glistening pirate of the streamside swoops occasionally to scoop up mosquitoes, gnats, and other small insects. Even when it alights on a leaf or a plant stem, the dragonfly is difficult to approach. The large eyes that cover most of its head see in nearly every direction at once, and its powerful wings are spread, ready for instant flight.

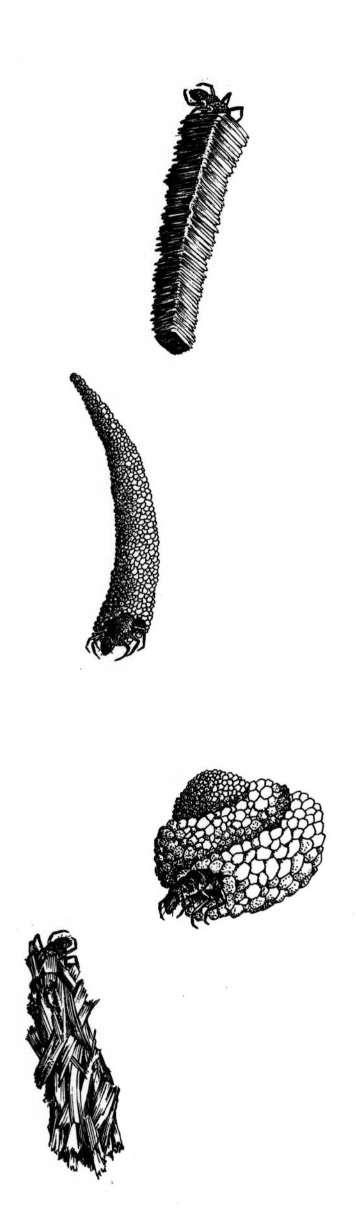

The precisely built cases of caddisfly larvae rank high among the engineering marvels of the insect world, with nearly every species characterized by its own distinctive building design. Larvae that live in fast-moving water usually build cases of sand grains and small pebbles, while those found in quiet pools tend to use lighter plant materials.

At the foot of the slope you pause beside a quiet, sunlit pool. The dark form of a fish—a trout, perhaps—darts through the water and disappears beneath an overhanging bank. A water strider skates across the surface, denting it but not breaking through, and casting weird shadows on the bottom. A water beetle with a silvery underside swims to the surface and takes on more air.

On the bottom of the pool a tiny bundle of twigs suddenly begins to move. If you look closely, you may see the head and legs of a "worm" projecting from the end of the carefully constructed case. This creature is the larva of a caddisfly, creeping about in search of a meal. Trout and other fishes hunt constantly for caddisfly larvae, but the insects' neatly camouflaged cases make them almost indistinguishable from the debris on the stream bottom.

Watch the pool and you may notice still other movement —a salamander hovering just above the bottom, or a water snake undulating by with its dark head held just above the surface. The quiet pool may have seemed almost empty when you first approached, yet it shelters these and many other inconspicuous animals whose lives depend on their ability to hide from ever-present enemies.

Life in the rapids

Upstream a bit, the water flows more rapidly, gurgling as it tumbles across the rocky stream bottom. You wade in and feel the current tugging at your ankles. Footing is precarious on the slippery rocks, but you make your way to the middle of the stream. Perhaps you wonder if any life can survive in the rushing water. What would the animals eat, and how would they manage to avoid being swept away by the relentless current?

Larvae of the caddisfly *Glossosoma* creep about on stream bottoms, building half-inch turtle-shaped cases of sand grains and tiny pebbles. Later they fasten their cases to current-washed rocks, often in crowded colonies, and remain there until their emergence as winged adult insects.

Actually, fast-flowing streams usually provide *habitats*, or places to live, for even more plants and animals than do quiet pools, although life in the rapids is more difficult to observe. Pick up a rock from the white water of a riffle, or rapid. As soon as you touch it, you are in contact with the basic life of the stream. The slimy surfaces that make the stream bed so treacherous for walking consist of a film of microscopic plants and bacteria, as well as one-celled animals that feed on them. Algae and other green plants use the energy in sunlight to manufacture the food that makes life possible for all animals in the stream.

Several long-legged, flattened or humpbacked insects undoubtedly are clambering over the slick surface of the rock. These are the young stages, or nymphs, of mayflies and stoneflies. There may also be caddisfly cases. Unlike the ones you discovered in the quiet pool, these are built of sand grains and tiny pebbles neatly cemented together.

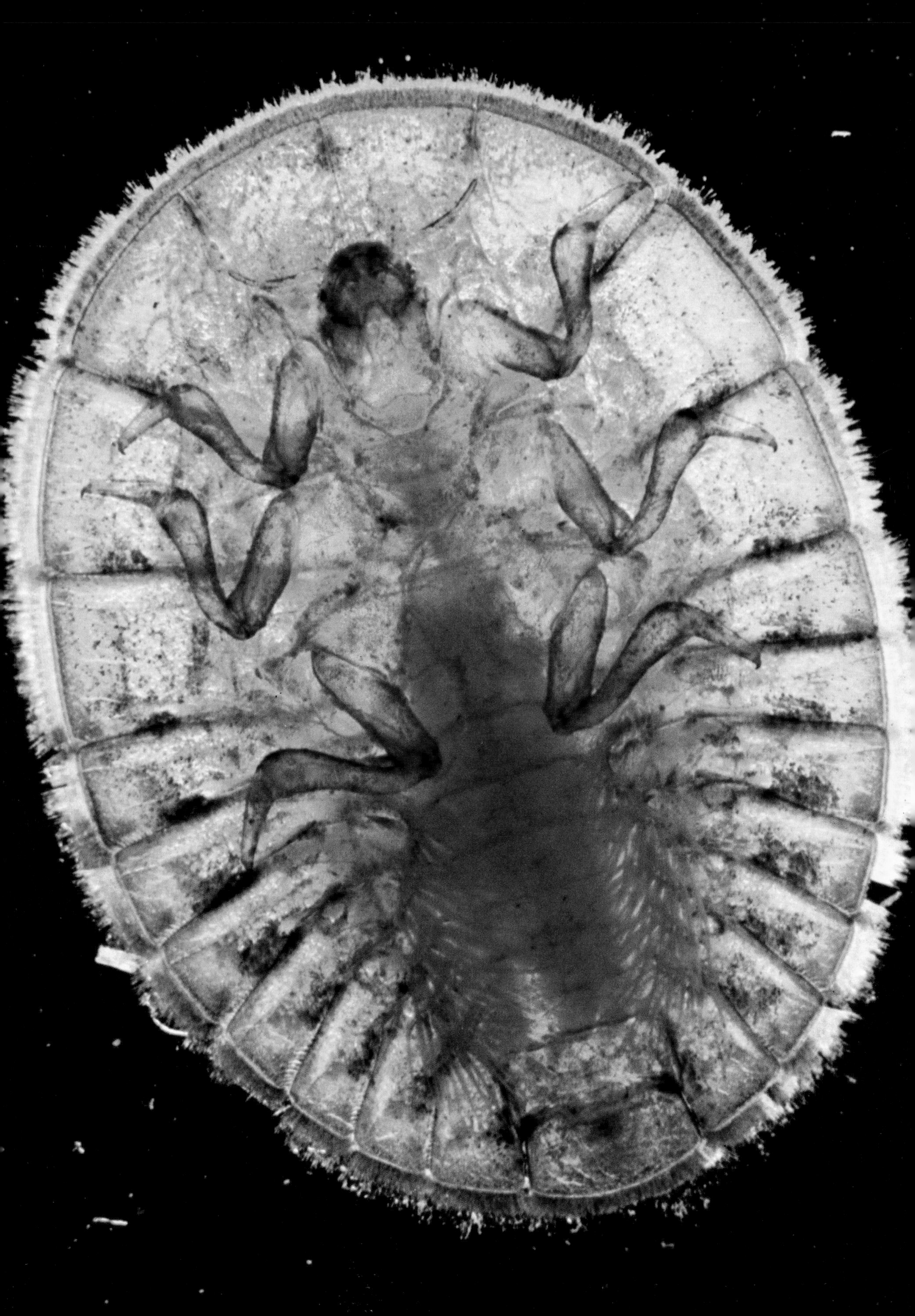

A water penny, the aquatic larva of an inconspicuous land-dwelling beetle, has been pried from a current-swept rock to reveal the head and legs that are normally hidden beneath its thin, flat shell. These living suction cups are so perfectly adapted to life in fast-flowing water that one species manages to survive at the very brink of Niagara Falls.

The paper-thin copper-colored creature clinging to the stone is a "water penny," the larva of a small beetle. Pry it off with your fingernail or a knife blade and look at the underside through a magnifying glass. You will see six tiny legs and the head and body of the insect, completely covered from above by its pennylike shell. Fastened securely to the rock by means of grasping legs and a suction disk formed by the margins of its streamlined shell, the water penny cannot be dislodged from its foothold on the stones even by fast-flowing water.

On the part of the rock that was exposed to the swiftest water, several half-inch-long black-fly larvae cling with hooks and suckers at the rear ends of their bodies, but now their heads are thrashing about in the air. In some streams, black-fly larvae are so abundant that their closely packed bodies look like sheets of dark moss on the rocks. Pairs of feathery brushes on their heads enable the insects to strain bits of food from passing water. You may also find seedlike black or brown cases scattered among the larvae. These are black-fly pupae. Inside each case, a wormlike immature insect is changing, or metamorphosing, to a flying adult. If the stone is still covered by water when the transformation is complete, the small, humpbacked adult fly nevertheless manages to emerge without wetting its delicate wings. Surrounded by a bubble of air, it simply floats to the surface and flies away.

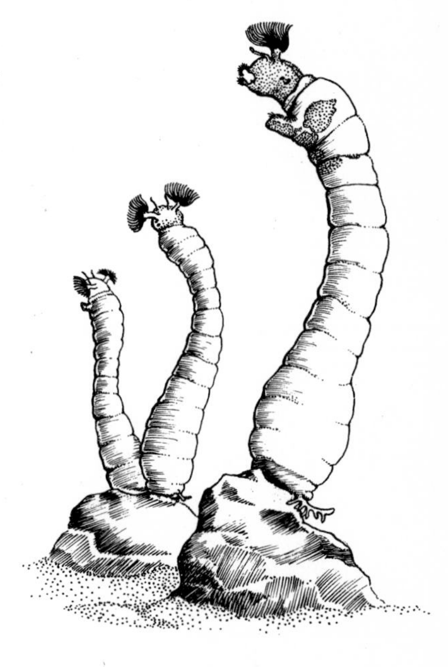

Half-inch-long black-fly larvae use tiny hooks to anchor themselves to rocks in swift-moving streams. Their fanlike head combs strain food morsels from the swirling current. If a larva should lose its grip, a silken life line keeps it from being swept away.

Occasionally you may find a stone that is practically covered by a soft green or yellowish crust. Although the growth vaguely resembles a fungus, it is not a plant at all, but a fresh-water sponge, one of the most primitive forms of animal life. Since it is fastened firmly to the rock, the sponge must depend on currents to bring it food. The surface of its body is covered with tiny pores that lead to open chambers and canals inside. Constant beating by minute hairs within the chambers keeps a stream of water flowing through the sponge's body and out through another set of slightly larger pores on the surface. Any bits of food carried in with the water are soon digested.

→

Wormlike black-fly larvae mingle with conical pupal cases.

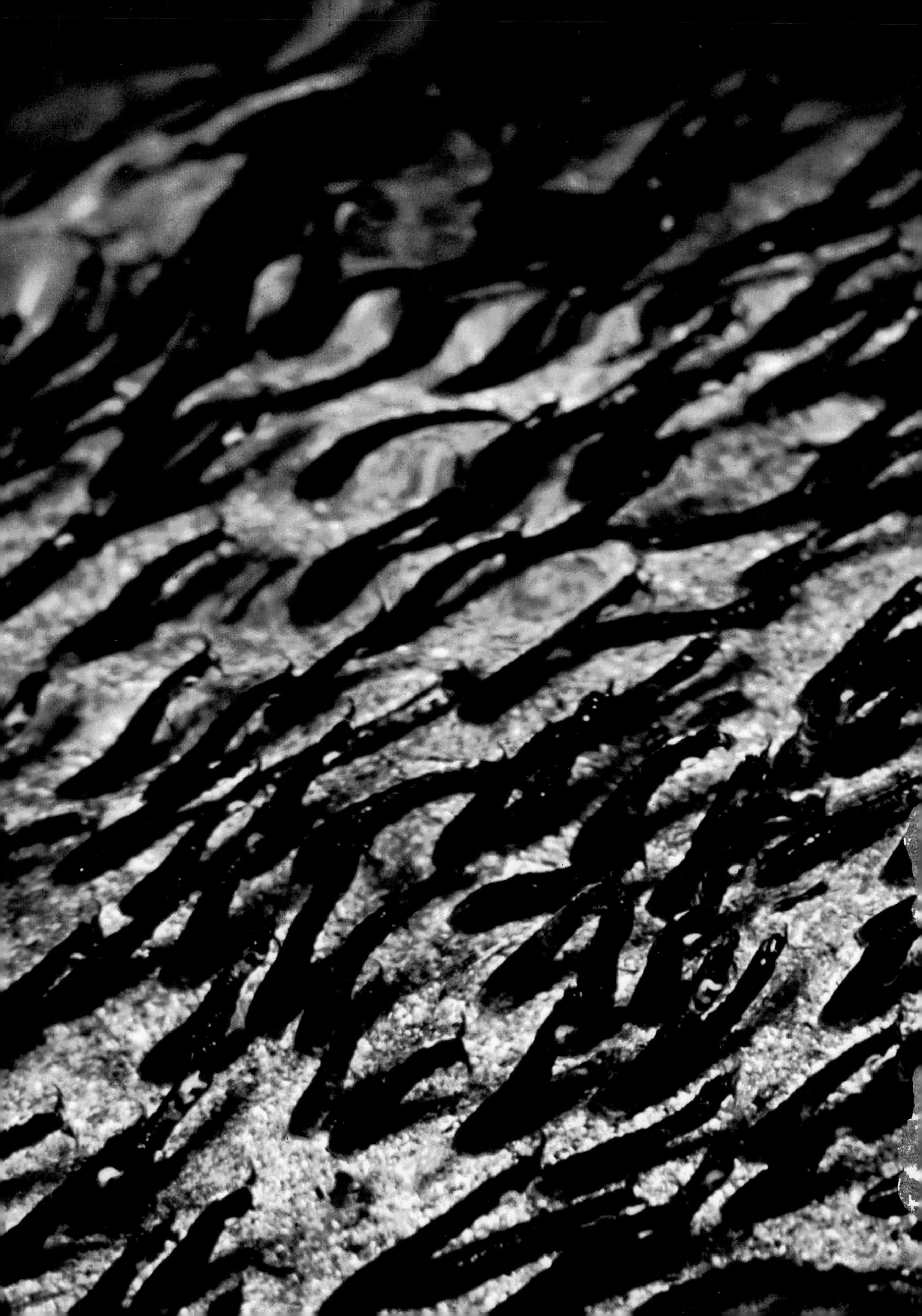

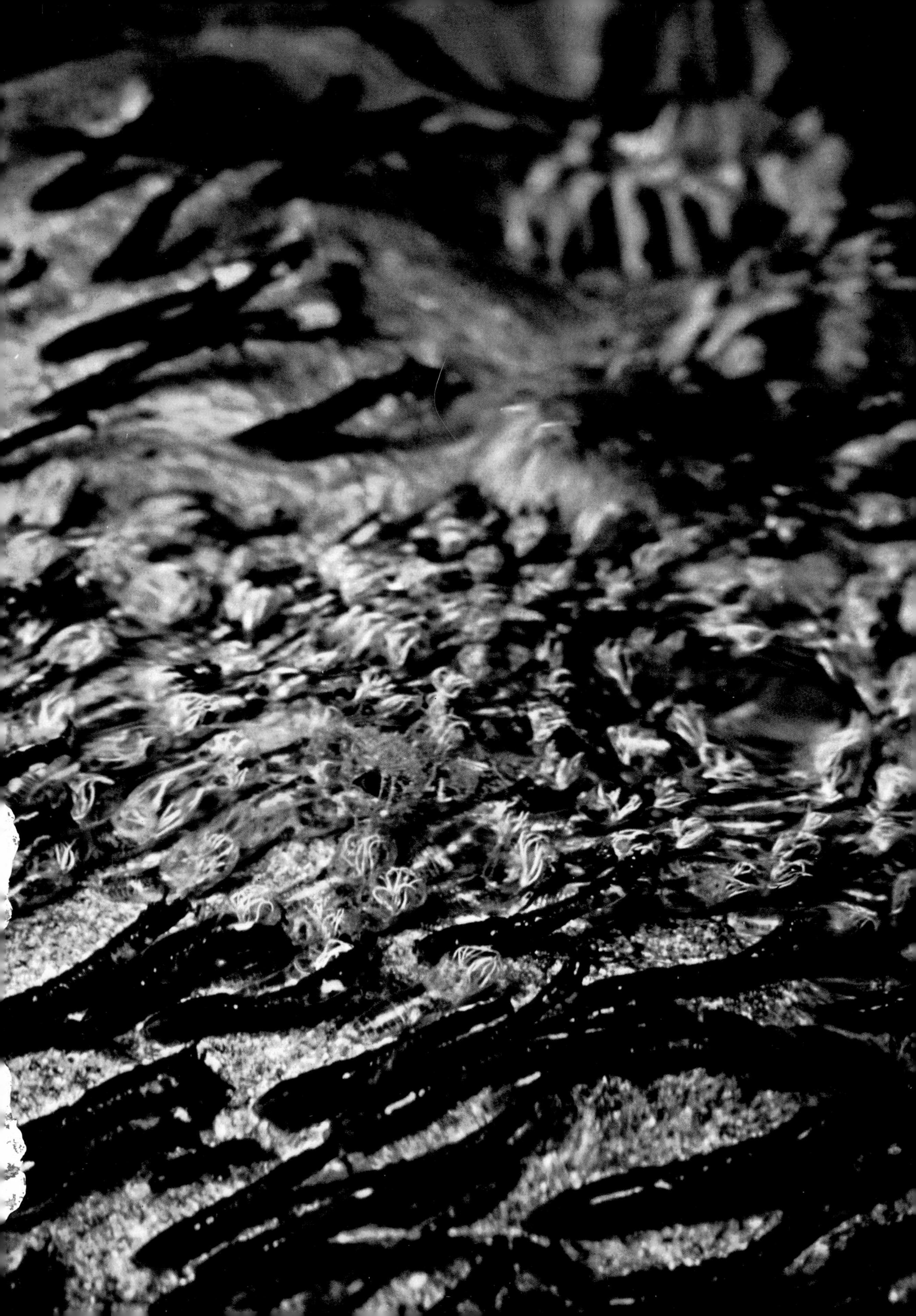

Many worlds in one

The stream, you discover, is more than just water on the move: it is water teeming with life, for the stream provides a seemingly endless variety of places where plants and animals can live. When it was lying in the stream, this single rock in your hand had an upper surface with water flowing over it, an undersurface touching or even stuck to the bottom, a front surface receiving the full force of the current, and a back surface on the protected, downstream side. Each of these surfaces provides a slightly different set of living conditions. As a result, each surface is inhabited by a slightly different community of plants and animals.

In addition, the rock may be smooth or rough, light or dark. It could have been in a permanently shaded or otherwise protected location. Other rocks may be partly exposed to air, or out of the water completely and moistened only by splashing. Shallow water with gravel or sandy bottom provides a very different habitat from deep water with large boulders. Fallen logs and rooted plants provide still other living places.

Moreover, each habitat within the stream is unstable and ever-changing. The water flows faster in spring and early summer, or even at the end of the day if its source is melting snow. After a heavy rain, flash floods may transform a placid

A stream is a dynamic, shifting habitat, subject to far greater seasonal changes than are larger bodies of water such as lakes. Heavy rains or melting snow, for example, can abruptly change this shallow sparkling stream . . .

stream into a muddy torrent that washes away much of the life and disrupts many of the habitats. Drought, on the other hand, may reduce a brimming brook to a feeble trickle that leaves many of the stream's inhabitants stranded on dry land. Others may suffer from overcrowding in the dwindling pools that remain.

Yet no matter what the conditions may be in any habitat within the stream, it is almost certain to be occupied by one or more kinds of life. Through millions of years of evolution, different plants and animals have developed a variety of adaptations for coping with the problems of survival in flowing water—how to fill their needs for food and oxygen, how to protect themselves from enemies, and how to reproduce their kind. Each form of life flourishes in the areas of the stream where its requirements for survival are best met.

Although no two of the thousands of streams in the United States are exactly alike, there is an overall similarity among them. The species of plants or animals occupying various habitats in the stream may differ from place to place, but the same general types are found in similar habitats almost everywhere. The general pattern of stream life is so consistent that a person familiar with the streams of New England or Virginia would very likely feel at home beside a stream in Colorado or California.

. . . into a swollen silt-laden torrent. In the face of these unstable living conditions, the plants and animals that inhabit streams have evolved a multitude of special adaptations that allow them to survive chaotic upheavals in their environment.

Sampling stream life

To get a closer look at some of the animals that live in a smoothly flowing section of the stream, hold a net or a piece of wire screen in the water, touching the bottom, and scuff the rocks on the upstream side of the net. Almost immediately, a bewildering variety of life washes into the net. Spread the net on a rock or, better still, dump your haul into a shallow white-enamel pan where the creeping, mostly dark-colored animals will be easy to observe.

If you are lucky, you may have netted a small bottom-dwelling fish, the sculpin. Most likely it was hiding beneath one of the stones you dislodged. It is seldom more than four or five inches long, with a head that seems too large for its tapering, scaleless body. The muddler, blob, or miller's-thumb, as it is also called, is mottled with dark patches that make it almost indistinguishable from the stream bed. Because its eyes are near the top of its head, the sculpin can easily scan its surroundings for passing insect prey as it hovers amidst the rubble. It maintains its position against the current by hanging onto stones with the large fanlike fins behind its head. Beautifully colored darters also hide among the stones in swift streams. If you watch one of these plump two- or three-inch-long fish, you soon will discover how it got its name: it darts forward with lightning-quick bursts of speed and stops as suddenly as it starts.

Two broomsticks and a piece of window screening make a simple but effective device for capturing stream animals. Overturning rocks on the upstream side dislodges an amazing variety of creatures, which the current then washes onto the screen.

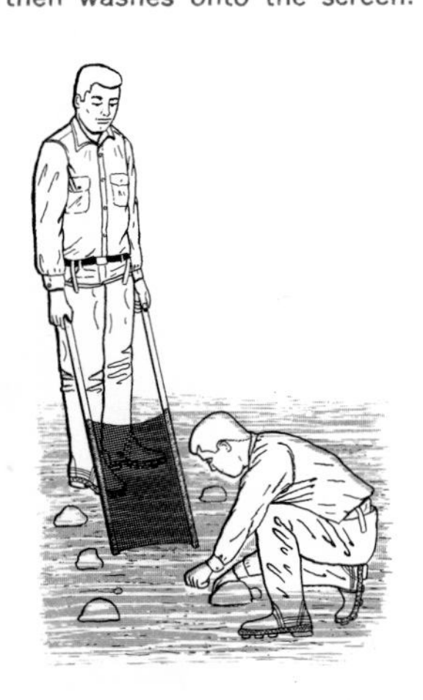

Another animal that lurks by day in spaces beneath stones is the crayfish. When alarmed, it can shoot backwards with surprising speed by flipping its tail beneath its body. If you should catch one, handle it with care; this miniature version of the lobster can inflict a painful nip with the fierce-looking pincers on its front legs. Although its claws are excellent tools for grasping insects and other small animals, this voracious scavenger of the stream bed usually is content to nibble on the remains of almost any dead plant or animal it may find.

Among the insects creeping across the bottom of the pan you undoubtedly will discover a number of stonefly nymphs. Six sprawling legs, each tipped by two stout claws, enable these inch-long insects to cling to the bottom in crevices between pebbles and under stones. A pair of taillike filaments projects from the rear of its body.

Most mayfly nymphs, in contrast, have three "tails" and only one claw on each leg. Although the flattened body form of many mayfly nymphs may seem to be an adaptation for

Crayfish are the best known fresh-water representatives of the group of animals including lobsters, crabs, and shrimp. These important members of the stream community eat practically anything—plant or animal, living or dead—that they happen to find. A wide variety of fishes, turtles, birds, and mammals in turn prey on these intriguing creatures, which also are known as "crawdads."

A wide variety of land-dwelling and flying insects pass their immature stages in streams. The stonefly nymph (*top*) feeds on algae and plant debris. Fringes of hairs on its legs help in swimming. The dobsonfly larva, or hellgrammite (*bottom*), is a fierce predator over two inches long. It hides beneath stones and among rotting leaves on the stream bottom.

resisting the force of the current, it is probably more important in permitting them to creep into protected crannies where they can escape the current altogether. Some species scrape algae from rock surfaces with their jaws, and others eat small animals in the stream. The nymphs also feed on dead leaves and other debris that falls into the water.

Probably the most ferocious-looking creature in your net will be a hellgrammite, the larva of the dobsonfly. Hellgrammites grow to two or more inches in length and have seven or eight pairs of fingerlike gills along their sides. If an unwary fisherman holds one carelessly when trying to bait his hook, he may get pinched by the strong, sometimes saberlike jaws. Adult dobsonflies are large black or brown insects that fold their wings over the body when at rest.

If the water where you netted your sample of stream life was not too fast-flowing, your haul may also include the nymphs of damselflies and dragonflies, another group of colorful streamside and lakeshore creatures that spend their immature stages in the water. These fierce predators, some-

times as much as two inches long, are among the most unusual of all hunting insects. Some species creep about in pursuit of prey, but others lurk amidst trash on the bottom and wait for passing victims. When an insect, tadpole, or small fish comes within reach, the nymph's hinged lower jaw suddenly shoots out a half inch or more and grasps the animal. With its prey impaled securely, the nymph then folds back its jaw and munches on its meal.

Damselfly nymphs, the more slender of the two types, have three platelike gills at the rear of their bodies. The stubbier dragonfly nymphs breathe by constantly pumping water in and out of the gills hidden within the tips of their bodies. Although they walk with perfect ease, dragonfly nymphs can also make a rapid escape by means of jet propulsion. You can demonstrate this by placing a drop of ink from a fountain pen just behind the tip of a nymph's body, then prodding the animal gently with the point. The ink will disperse into a cloud as the nymph moves forward by expelling a jet of water.

The damselfly nymph is easily recognized by the three leaf-shaped gills at the tip of its abdomen. On the closely related dragonfly nymph, the gills are internal. The four wing lobes behind its head are reminders that the inch-and-a-half-long damselfly nymph someday will emerge from the stream as a flying adult insect.

Torrents and waterfalls

Although beautiful to look at, the foaming, roaring sheets of water that tumble over cliffs and rock ledges seem the least likely places of all to look for life along a stream. What creatures could possibly survive in such surroundings?

Within the raging water of the torrent itself, there are no permanent residents. But look more closely and you will probably find the tiny larvae of net-winged midges clinging to rocks practically under the main part of the falls. Scarcely over a quarter of an inch long, each larva is divided into seven lobes, so that it resembles a miniature string of beads. If you examine it with a magnifying glass, you will discover that it is equipped with a row of the most efficient holdfast organs in the animal kingdom.

On the underside of all but one of the body sections, the larva has a firm ring scored with radiating grooves. At the center is a plug attached to muscles in its body. When the muscles are relaxed, the plug is drawn upward, creating a vacuum within the suction ring that holds the larva in place. Cement glands and waterproof hairs around the sucker complete the device.

Since three suckers are enough to keep the insect from being washed away, it frequently detaches three at one end of the body, swings around, attaches them elsewhere, and then detaches the other end. In this ungainly fashion, it moves fairly rapidly across rocks in spite of the falling water. After living for several months in the waterfall, the midge finally transforms into an inconspicuous mosquito-like adult that hovers over the falls to mate. The female is believed to dive directly into the torrent to lay its eggs.

Another insect found in riffles and near waterfalls is the net-building caddisfly larva called *Hydropsyche*. By spinning perfect silken nets attached to stones, this wormlike creature traps small animals carried downstream by the rushing water. The larva creeps out periodically to harvest the feast snared in its net and then retreats to a crevice in the rocks.

Here by the falls and rapids is also a good place to look

No portion of the stream is without life; even the waterfall has its characteristic inhabitants. Net-winged midge larvae, for example, cling to water-swept stones, while larvae of the caddisfly, *Hydropsyche*, stretch silken nets across the current to snare bits of food.

YOSEMITE NATIONAL PARK

Plummeting a total of 2425 feet—the equivalent of more than a dozen Niagaras—Yosemite Falls (*right*) is the best known of Yosemite National Park's attractions, but it is by no means the only one. A number of lesser but still spectacular waterfalls are scattered through the park, together with more than 200 lakes. The Merced River (*left*), winding lazily through Yosemite Valley, provides excellent boating, fishing, and swimming. For hikers and campers, there are over 750 miles of trails, while the less adventurous can enjoy outings on some 200 miles of scenic roads. The park also includes several of the finest remaining stands of giant sequoias, an abundance of birds and mammals, and impressively rugged valleys and peaks scarred by ancient glaciers—all within a few hours' drive of busy San Francisco.

for birds. If you live in the West, you may catch sight of the dipper, or water ouzel, a chunky blue-gray bird with a tinge of brown on its head and shoulders. The starling-sized dipper lives almost exclusively near swift-flowing mountain streams. As it hops from rock to moss-capped rock projecting from the rapids, it seems a busy but fairly drab bird. And then it disappears from sight. You blink in disbelief when you realize that the bird has plunged directly into the torrent. With perfect ease, the dipper defies the current and walks across rocks on the bottom, foraging for caddisfly larvae, stonefly nymphs, and other insects. In deeper water it swims to the bottom by flapping its stubby wings. Even in the dead of winter, when streams are fringed with ice, the dipper braves the roaring current and plunges in to feed. Its song, a series of bubbling trills alternating with rich flutelike notes, is familiar to many mountain climbers.

Look for the dipper's nest in crevices on dripping moss-covered ledges along the stream, or even directly behind the sheets of water pouring over the falls. The ovenlike mound of plant material, usually green and yellow mosses, is about a foot in diameter, with an opening at the side or near the bottom. It may be difficult to locate, since the mosses woven into the nest generally continue to grow in the fine spray that splashes from the waterfall.

Undaunted by the swift currents and near-freezing temperatures of mountain streams, the dipper feeds by plunging directly into the water. As casually as a robin hunting worms on a suburban lawn, it scurries across the flooded stream bed and probes for insects hidden among the rocks.

Spotted sandpipers are more at home in water than you might suppose from the appearance of this bedraggled youngster. Like dippers, spotted sandpipers are accomplished swimmers and sometimes walk along the stream bottom, foraging for insects, crustaceans, and small fish.

Life along the bank

In the East, a common bird of streamsides is the water-thrush, an olive-brown warbler with brown-streaked breast and a white or yellow line over each eye. It is a common sight beside woodland streams, bobbing up and down as it searches for insects among damp mosses and the snarled debris of rotting leaves and twigs. Its delicate nest is usually concealed among exposed tree roots that dangle beneath the shelter of an overhanging bank.

The spotted sandpiper also frequents the banks of streams, or almost any other body of water, across most of the United States. It is gray-green with large, dark polka dots on its white breast. Scurrying across fallen logs or darting along the bank, the comical sandpiper teeters up and down like a mechanical toy. Occasionally it even dives into water from above, and, like the dipper, hunts insects and other prey on the bottom.

If you look closely, you may discover toad bugs near the

water's edge, although their flattened oval bodies almost match the color of sand or mud. With their weirdly protruding eyes, they look quite toadlike as they squat on the sand and wait patiently for smaller insects to come by. A sudden lunge, and the prey is captured.

Still higher on sandy banks you may see small holes about a quarter of an inch across. Probe with a twig and you may come up with the larva of a tiger beetle. When undisturbed, the larva lies in ambush just inside the entrance to its home, waiting for an unsuspecting ant or some other insect to approach. In a flash the larva reaches out from its hole and grabs its victim with strong, curved jaws.

As adults, tiger beetles are brown or greenish metallic-looking insects spotted with irregular pale markings. As you approach a sandy area along the bank, you may see a few of the wary half-inch beetles standing over the hot sand, ready to take off at the slightest alarm. Although tiger beetles are among the fastest-moving of all insects, you may be able to catch one if you approach it cautiously with a net.

Where the stream is bordered by damp mud, the tracks of larger streamside visitors dot the surface, each one as distinctive as a signature. Many animals come regularly to the stream for food or water. But since they usually slip down to the edge of the stream under cover of darkness, you must rely on an occasional stroke of luck to glimpse the animals. Yet, with a little practice, almost anyone can learn to recognize the tracks of each and unravel the fascinating stories written in their trails.

There are almost certain to be pairs of delicately pointed imprints of deer hoofs in the mud. In Yellowstone National Park, the Great Smoky Mountains, or other wild areas, you may see the broad-soled, five-toed tracks of bears. In the

In a face-to-face view, the toad bug is a fearsome sight. But when seen from above, the flattened half-inch-long insect is nearly invisible against the speckled sand or mud of the stream banks where it lives.

Pacific Northwest, especially, black bears often gather along the shores of streams and rivers to gorge themselves on schools of salmon migrating upstream to spawn.

A curious track with an awkwardly turned-back thumbprint marks the rambles of the ungainly opossum, while a strangely human-looking five-fingered imprint is the paw mark of the raccoon. The black-masked, bushy-tailed raccoon eats almost anything, both plant and animal. When it visits the stream, it is usually intent on capturing a meal of aquatic insects, crayfish, clams, and even fish. Although the raccoon frequently appears to be washing its dinner in the water, it is more probably trying to soften hard food. Or possibly it is simply examining its catch with its nimble, sensitive fingers.

Near large streams you may find the tracks of muskrats, otters, minks—the possibilities are almost unlimited. Along smaller streams, you'll need no tracks to tell you when you have entered beaver territory; conspicuous dams will be evidence enough of their presence. These sturdy structures, built of branches, stones, and mud, create deep ponds where the

This doe's big black-fringed ears mark her as a mule deer, the most common species of the western United States. When the need arises, these frequent visitors to streams and water holes prove themselves skillful swimmers.

Virtually any river or stream in America is likely to be visited regularly by raccoons. Despite persistent hunting and trapping by man and his relentless destruction of wilderness areas, these intelligent roly-poly mammals continue to flourish throughout most of the United States. Although raccoons are able swimmers, they do not dive for their food; they prefer to squat on shore or splash through the shallows and snatch up anything edible with their deft forepaws. Female raccoons bear three to six young in the early spring. The kits stay with their mother throughout the summer, learning how to find food and avoid enemies, and by late fall they are on their own.

beavers build domed, islandlike lodges. Hiding by day in the dim but safe interior chamber of the lodge, the beavers leave through underwater tunnels after dark and swim to shore. There they gnaw the trunks of aspens and willows, to feast on the tender inner bark. In late summer and autumn, they drag many branches back to the pond and anchor them in mud near the lodge. Throughout the winter, when the surface of the pond is locked beneath a layer of ice, they feed on this handy supply of bark and twigs.

Neatly adapted to life in the water, the beaver's large hind feet are webbed like a duck's. Besides making its feet efficient paddles for swimming, the webbing helps support the heavy animal when it walks on soft mud. (A four-foot-long adult may weigh forty or fifty pounds!) The scaly paddlelike tail acts both as a rudder for swimming and as a prop when the beaver is at work cutting a tree. Coarse, well-oiled hairs projecting above dense, soft underfur keep the beaver dry when it swims. The beaver, in fact, is waterproof in almost every way. Flaplike valves close over both ears and nostrils whenever the beaver plunges into water. Since its lips can close behind the stout chisellike teeth at the front

Heavily webbed hind feet and a flat scaly tail make the beaver unmistakable among the stream's mammal residents. At one time brought close to extinction by fur trappers, this largest of American rodents is now protected by conservation laws.

of its jaws, the beaver does not swallow water when it gnaws on food underwater.

Probably the most astonishing of all aquatic mammals is the mouselike water shrew. If you catch sight of a tiny animal seeming to stand on its head in the shallows of a beaver pond or some other quiet pool, it is almost certain to be this resident of stream borders and pond margins. The water shrew regularly enters water and dives to the bottom to probe for insects and other food hidden among the sand and pebbles. The shrew's appetite is so voracious that it eats several times its weight in food each day.

The shrew moves constantly in and out of water, yet it never gets wet. Its fur is so fine that it does more than merely repel water. Each time the shrew submerges, a film of air clings to the tips of the hairs. As a result, its glossy, dark gray body looks silvery underwater. This film of air lends so much buoyancy that the shrew can bob easily to the surface and swim to shore, where, like a miniature dog, it shakes its tiny body dry. Fringes of stiff hairs on its hind feet form efficient paddles, while another row of bristles on its tail seems to serve as a rudder.

One of the rarest and certainly the tiniest of aquatic mammals is the water shrew. Although an adult beaver may top fifty pounds, a full-grown water shrew usually weighs less than an ounce. This little creature is an able swimmer and also can walk on the stream bottom in search of insect prey.

With its six legs dimpling but not breaking through the surface film, the water strider skates with surprising speed and agility across slow streams and ponds. Its middle legs do most of the "rowing." The hind legs act as steering rudders, while the forelegs are used for grasping prey, mostly insects smaller than itself.

Life on the surface film

Occasionally the water shrew accomplishes the impossible, or so it seems. With its feet flailing vigorously, it skitters for short distances across the very surface of the water. Seek out a quiet pool or backwater and you will find still other plants and animals living on this seemingly insubstantial boundary line between air and water.

Tiny duckweeds often dot the surface of still waters with living flecks of green. In the South there may be velvety mats of the delicate water fern *Azolla* or lush expanses of showy water hyacinths. All drift freely on the surface. Other plants, such as water lilies, are rooted in the bottom but have leaves that float like rafts. They are anchored by flexible stalks that permit the leaves to rise and fall with changes in the water level.

More often than not, the leaves of floating plants are honeycombed with airspaces that make the leaves buoyant enough to drift on the surface. But what about the slender, long-legged water strider that skims so nimbly across the water in search of insects floating on the surface? What keeps it from sinking?

To get some idea of how the water strider manages this feat, try floating a clean needle on still water in a saucer. The surface of the water bends down under the weight of the needle, but the needle remains afloat. This happens because the water molecules of the surface have a stronger attraction for each other than they do for the air above. The result is an unbelievably thin but sturdy film, or "skin," on the surface. This attraction of water molecules is called *surface tension*.

The water strider takes advantage of surface tension by distributing its weight widely across the film with long, sprawling legs. In addition, the tips of its legs are fringed with waxy hairs that help spread its weight and repel the water. The strange shadows cast on the bottom beneath a water strider result from the slight dimples its legs make in the surface film.

Movement also helps in staying on the surface, as any water skier knows. By moving rapidly, the heavy-bodied water spider and the nimble water shrew manage to scoot across the surface without sinking. Tiny springtails may also

In quiet water, even creatures as large as snails can take advantage of the surface film. But instead of walking on top of the water as the water strider does, the snail uses its fleshy "foot" to glide along the undersurface of the film.

throng on the surface of quiet coves, and in almost any body of water you are likely to find whirligig beetles. Like shiny, animated watermelon seeds, they gyrate crazily on the surface, yet never seem to bump each other.

Although its underside breaks the surface film, the whirligig beetle's back is waxy and remains dry. Catch one and you will see how well adapted it is to its strange air-and-water environment. Each eye is divided into two parts, an upper part on top of its head for seeing in air, and another part on the underside for seeing in water. This creature of the surface film, in fact, enjoys the best of two worlds. It is equally adept at diving beneath the surface and flying through the air.

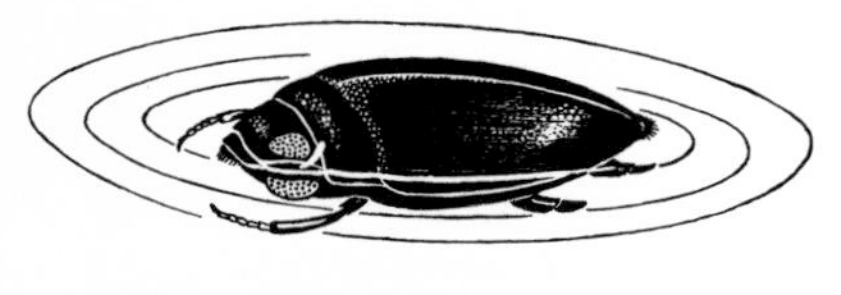

Natural "bifocals" — eyes divided into separate upper and lower halves — permit the whirligig beetle to see both up and down as it zigzags across the surface film of water in quiet pools.

A few animals even manage to survive on the underside of the surface film. Fresh-water snails sometimes creep upside down beneath the film, with their broad, fleshy "foot" moving along as though the surface were a solid glass plate. Larvae of many mosquitoes also spend a great deal of time resting just beneath the surface. In order to breathe, they poke tiny water-repellent snorkels, or breathing tubes, through the surface film.

Breathing underwater

Nearly all living things require oxygen. The simplest forms, such as one-celled plants and animals, absorb it directly from water or air through their body surfaces. More complex forms need more oxygen and are equipped with special structures for obtaining it. There is always plenty of oxygen available for land animals like ourselves. Nearly twenty-one percent of the air we breathe is oxygen, which enters our systems by way of the lungs. These are basically a series of moist membranes underlain by a network of ultrafine blood vessels. Oxygen passes through the membranes and into the blood, where *hemoglobin*, a complex substance responsible for the red color of blood, carries it to all parts of the body. Carbon dioxide, the gaseous waste product of our life processes, passes out of the blood surging through the lungs and is expelled from the body when we exhale.

Land-dwelling insects breathe in a different way. Catch a large grasshopper and examine it closely. Along the sides of its body, from wings to tip, you will see a row of tiny holes. Each one is the opening to a minute tube which

branches into hairlike tubes within the body. Air reaches all parts of the grasshopper's small body through these fine tubes, or *tracheae*.

Water poses special problems in breathing, and the solutions are varied. Some animals, such as beavers and otters, surface occasionally for a gulp of air. A beaver can remain submerged for fifteen minutes or so. Its extra-large lungs hold a good supply of air, and an additional reserve of oxygen is stored in the blood in its unusually large liver.

However, most aquatic animals use the oxygen dissolved in water. Watch a goldfish in an aquarium. It appears to be gulping water, but actually the water that enters its mouth is passing over its gills and leaving through slits behind its gill covers, the stiff flaps at the sides of its head. If you catch a goldfish or a sculpin and gently lift one of the gill covers, you will see layer upon layer of delicate, bright red filaments. These are gills. As water passes over them, it gives up oxygen and carries away carbon dioxide that is brought very near the surface of the gills by a rich supply of blood.

Like all fish, this largemouth bass "breathes" by continuously pumping water in through its mouth, over its gills, and out through openings just behind its head. This action supplies the fish's blood with oxygen and relieves it of carbon dioxide. Rows of fringelike gills extend back from the series of bright red supporting arches visible in the throat of this bass.

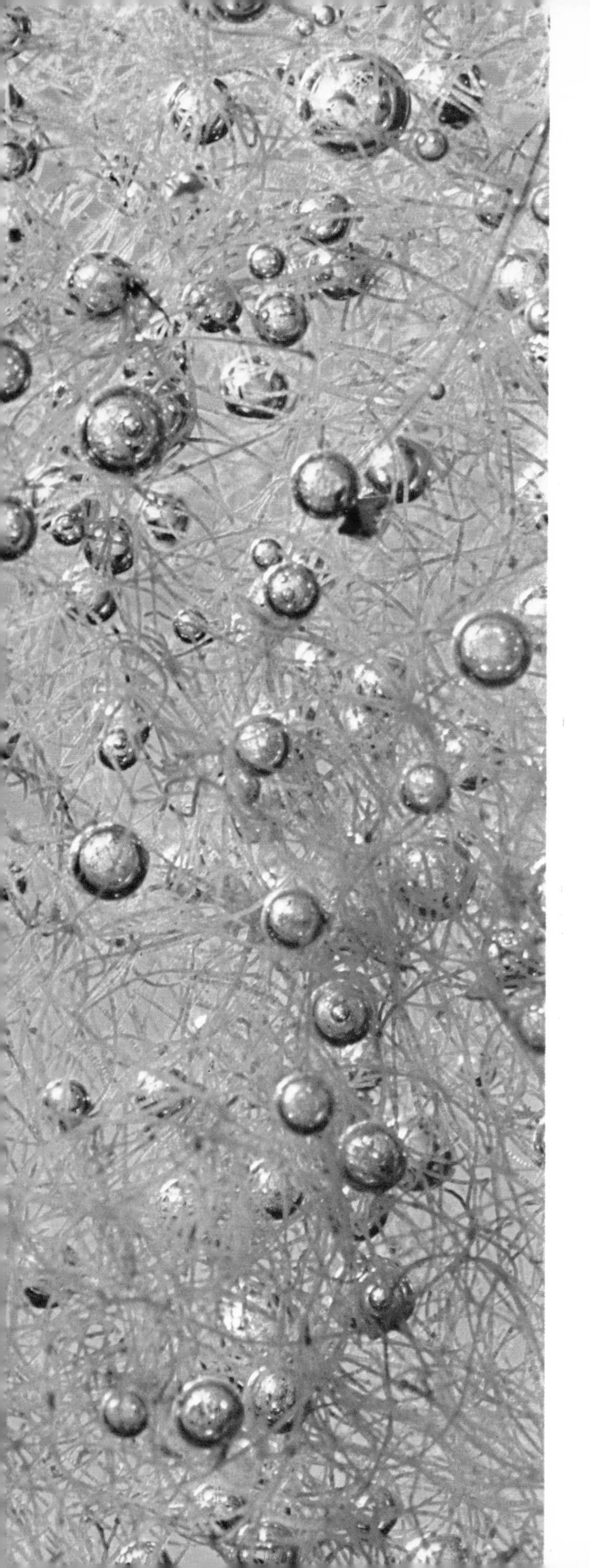

THE PROBLEM OF OXYGEN

Nearly all plants and animals, regardless of where they dwell, have in common certain basic life necessities, among them food, water, and oxygen. For land dwellers such as man, water tends to be the most crucial of these, for it is often in short supply. We live in an atmosphere that is one-fifth oxygen, more than enough to meet our normal requirements. But in a stream the situation is reversed: water is abundant, but oxygen often is difficult to obtain.

Some stream creatures solve the problem by rising to the surface and breathing atmospheric oxygen. This group includes many of the insects, some of the snails, most of the amphibians, and all turtles and mammals. But many others, from one-celled protozoans to the biggest fish, rely entirely upon the oxygen dissolved in water and depend upon its continual replenishment.

The atmosphere is one source of replenishment: wherever air and water meet and mix, oxygen is abundant. Fast cascading streams like the one on the right are rich in oxygen. Slow-moving stagnant waters are oxygen-poor.

A less obvious but equally important source of oxygen is the stream's green plants, especially tiny algae like those on the left. In the process of manufacturing food, these green plants release a great deal of free oxygen into water. Later in this book you will see how green plants form the basic foodstuff of the stream community; here you see their second great contribution—helping to solve the oxygen problem.

Unlike most salamanders, the mudpuppy remains a gill breather all its life. Mudpuppies are found in a variety of aquatic habitats, and the condition of an individual's gills will tell you something about the water in which it lives. If the water is poor in oxygen, the gills will be large and bushy. If the water is oxygen-rich, the gills will be smaller and less conspicuous.

The gills of a salamander larva function in much the same way, except that they are external, appearing as branching or feathery growths on each side of the neck. In most species, the gills disappear as the larva transforms into a land-dwelling adult. The frog tadpole also has external gills during its earliest stages, but later the gills are covered by a flap of skin. The tadpole gulps water with its mouth, passes it over the gills, and then expels it through a small tube, the *spiracle*, on the left side of the body. You can watch a tadpole breathing if you put one in a pan of water, then place a drop of ink in front of its mouth; as the tadpole gulps the ink, a stream of tinted water is expelled through the spiracle. In the adult frog, the gills are replaced by air-breathing lungs. Although the frog absorbs some oxygen directly through its skin, it cannot remain underwater indefinitely but must return occasionally to the surface.

The beetle and the bubble

Aquatic insects have developed a variety of amazing devices for tapping surface air. The rat-tailed maggot, for example, dwells in shallow, stagnating pools and decaying muck. At the tip of its half-inch body is a slender telescopic tube that can be extended an inch and a half or more to take in surface air.

The larvae of *Donacia* beetles have evolved an even more ingenious solution to the problem of getting air. The metallic-looking adults lay eggs on floating lily pads in ponds and backwaters. When the larvae hatch, they drop to the bottom and seek out water-lily stems. By rasping a hole in the stem, the beetle larva taps the oxygen stored in airspaces in the plant stalk. With the tip of its body thrust in like a plug, the insect breathes air brought to it by the plant's unusual air system.

Water beetles, water boatmen, and several other insects, on the other hand, carry surface air underwater on their bodies. Catch a water beetle and put it in some clear water. When you hold the beetle in air, its undersurface is dull. Yet when you plunge the beetle underwater, a silvery film coats its underside. This film is a bubble of air that the beetle carried underwater. The air is held to the beetle's body by

A glistening bubble of surface air adheres to a water boatman's abdomen as it dives toward the bottom of a pool. The agile boatman usually remains submerged for fifteen minutes or so before returning to the surface to replenish this portable oxygen supply.

a dense coating of waxy water-repellent hairs that are visible through a microscope or a good magnifying glass.

While the insect is underwater, oxygen enters its system of air tubes, or tracheae, and is gradually used up. As the oxygen is consumed, an oxygen deficiency develops in the bubble. Since gases tend to reach an equilibrium, dissolved oxygen then passes into the bubble from the surrounding water. Thus the film of air itself acts as a kind of gill, permitting the beetle to stay underwater for a longer time than it could if it depended only on the oxygen originally in the bubble. Under normal conditions, however, the beetle refreshes the bubble every few minutes by breaking through the surface film.

But bobbing periodically to the surface can be dangerous, especially in running water. A small insect may be swept downstream before it can return to the bottom. Thus, it is hardly surprising that a good many immature insects depend solely on the oxygen dissolved in the water.

Insect gills

Some insects, especially larval forms, absorb oxygen directly through their body walls. Others, such as certain of the wormlike midge larvae, known as bloodworms, possess so-called blood gills. These fleshy protuberances on their bodies are permeated by red blood, which, like human blood, contains hemoglobin. But the great majority of gill-bearing insects have developed modifications of the more usual insect breathing mechanism, a network of tracheae throughout their bodies.

More than any other group of aquatic animals, the insects have evolved a fascinating array of adaptations for breathing. The *Donacia* larva, for example, gets oxygen by burrowing into air-filled roots of water lilies and other plants. The water scorpion obtains surface air through a tube formed by two tail filaments edged with interlocking bristles. The water caterpillar absorbs dissolved oxygen through tufts of fingerlike tracheal gills, while many mayfly nymphs bear rows of flattened, platelike tracheal gills. The adult diving beetle *Dytiscus* carries a bubble of air beneath its wing covers; as a larva, it bobs periodically to the surface and takes air directly into its tracheal system.

Donacia BEETLE LARVA

WATER SCORPION

WATER CATERPILLAR

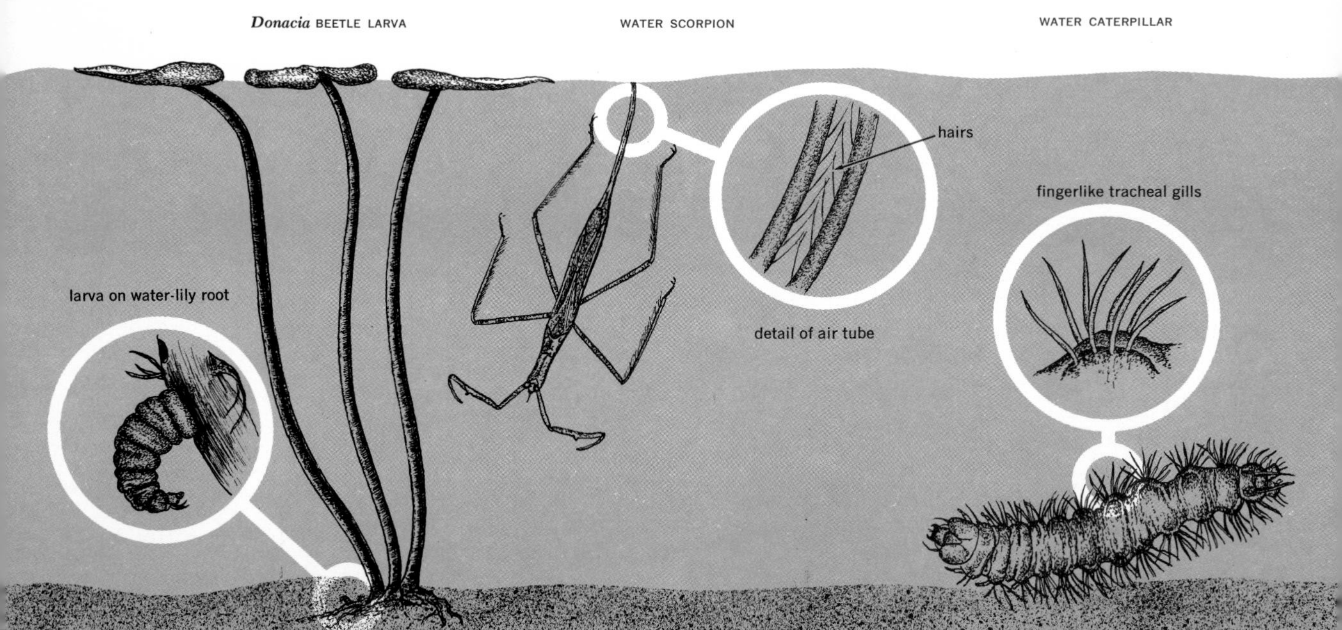

Look at a damselfly nymph. Three delicate leaflike structures project from the tip of its tail. A magnifying lens will reveal an intricate tracery of minute branching veins in each of these tracheal gills. Instead of opening to the surface, however, the ends of the tubes are sealed. The walls of the gills are so thin that oxygen passes readily into the air tubes and is distributed throughout the creature's body.

Many species of mayfly nymphs have similar gills, except that instead of being at the tip of the tail, the thin paddle-shaped gills lie in a row along either side of the body. On stonefly nymphs, water-beetle larvae, and many other insects, the gills are simple fingerlike projections or elaborately branched featherlike structures on the undersurface or at the sides of their bodies. The constant flow of water across their surfaces keeps the insects amply supplied with oxygen.

In some cases, the insects may create currents of water across their gills. Caddisfly larvae, for example, have a special problem. Their elaborate cases are fine for camouflage but make breathing a bit difficult. If you remove a larva from its case, you will discover that its gills are simple threadlike structures along its sides. Place the larva in a pan of water, along with a supply of fine chips of broken glass. The larva will immediately seek protection by building a new case. Since the only available material will be the bits of glass, its new case will be more or less transparent, allowing you to observe what is happening inside. Once the caddisfly larva is settled in its new home, you will notice that its body is frequently in motion, undulating up and down to keep a current of water flowing in and out of the case. The less dissolved oxygen the water contains, the more vigorous the movements will become.

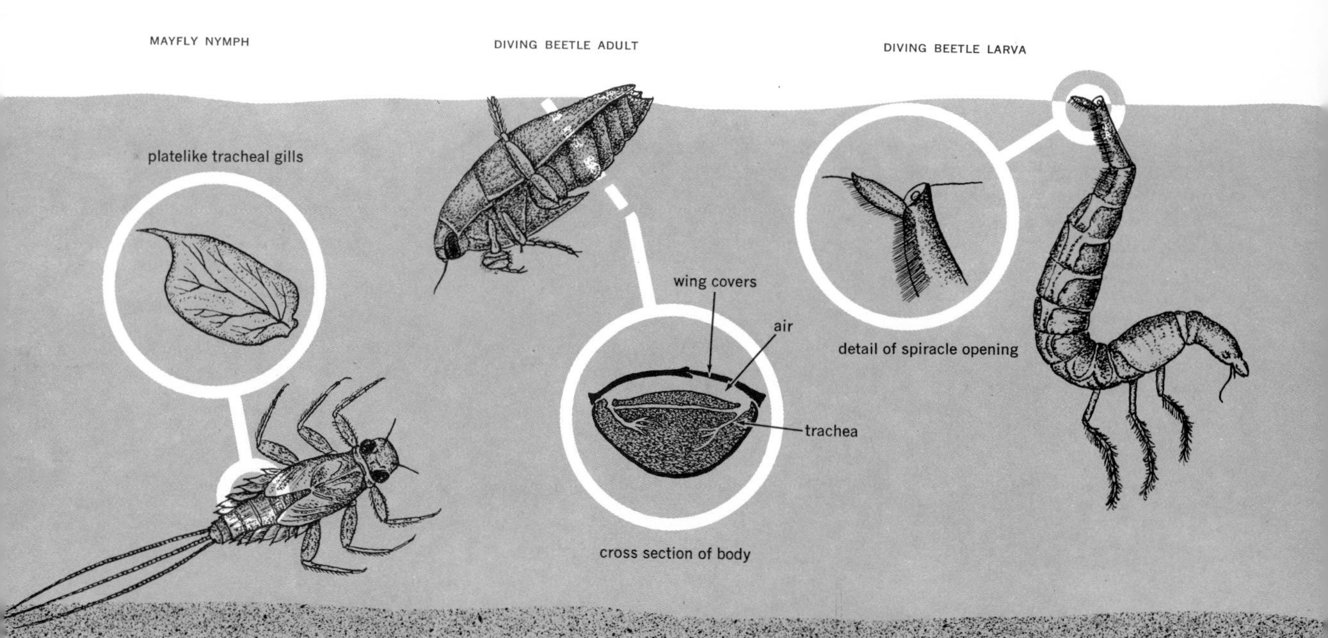

SURVIVAL IN SWIFT WATER

Fast-flowing streams present a unique set of hazards to creatures that make their homes there. In the face of the never-ending downstream tug of currents, the most basic daily activities can involve formidable problems. Even the simple matter of staying in one place presents a challenge. All the essential processes of life — seeking out food, avoiding enemies, finding a mate, or bringing forth offspring — would seem almost impossibly difficult under such conditions.

Yet living organisms are supremely adaptable. An astonishing variety of animals have evolved specialized structures and habits that enable them not only to cope with the hazards of life in swift water, but also to take advantage of its benefits. For such an environment does offer advantages as well as drawbacks. Indeed, many of the creatures that have come to flourish in strong currents are so well adapted to life there that they quickly perish when transplanted to quieter portions of the stream.

The sculpin, a big-headed six-inch-long fish common in cold upland streams, resists the pull of the current by bracing its oversized forefins against the upstream sides of stones. A weak swimmer, the sculpin hides during the day and emerges at night to clamber among stream-bed rocks in search of prey.

The crayfish avoids the current by hiding beneath sheltering rocks and stones on the stream bottom. Normally it uses its legs for locomotion as it creeps warily from niche to niche. If forced into open water, it flips its powerful tail forward in order to move backwards with surprising speed.

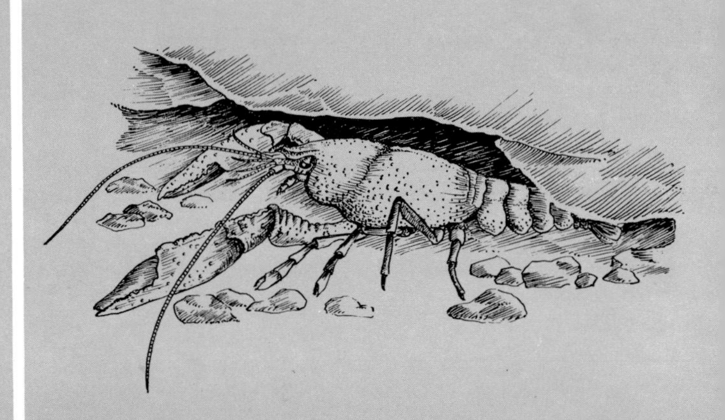

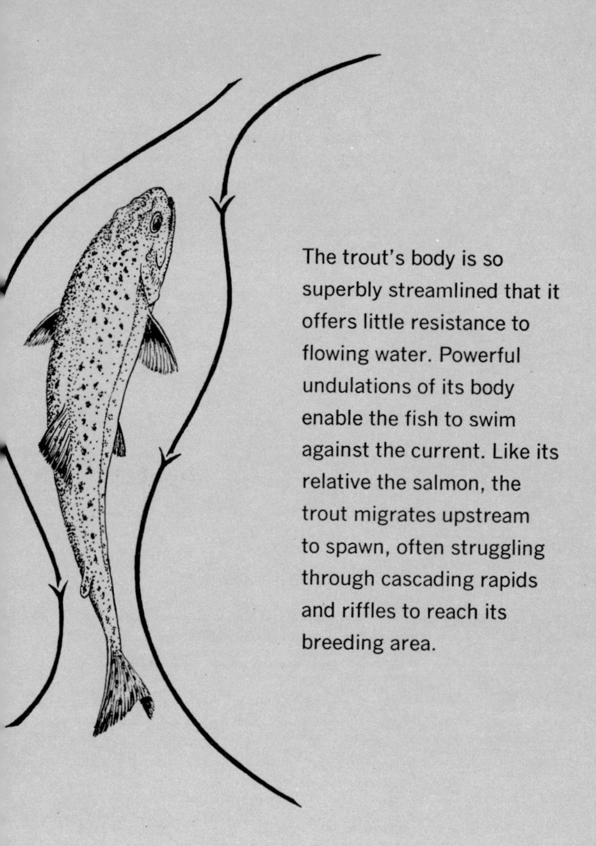

The trout's body is so superbly streamlined that it offers little resistance to flowing water. Powerful undulations of its body enable the fish to swim against the current. Like its relative the salmon, the trout migrates upstream to spawn, often struggling through cascading rapids and riffles to reach its breeding area.

The mayfly nymph, *Isonychia,* relies on swift currents to bring it food. Clinging to a stone, it spreads its bristle-studded front legs to net food particles being swept downstream. Most mayfly nymphs, like the one in the background, graze on the film of algae that covers rocks.

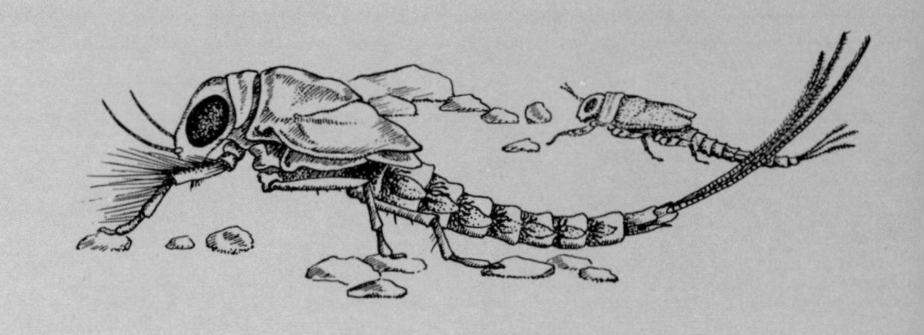

The larva of the net-winged midge thrives in torrents and waterfalls, where it defies the current by clinging to rocks with six suction disks that form a row along its underside. Creeping slowly from place to place, the quarter-inch-long larva feeds by scraping microscopic plants from the rocks.

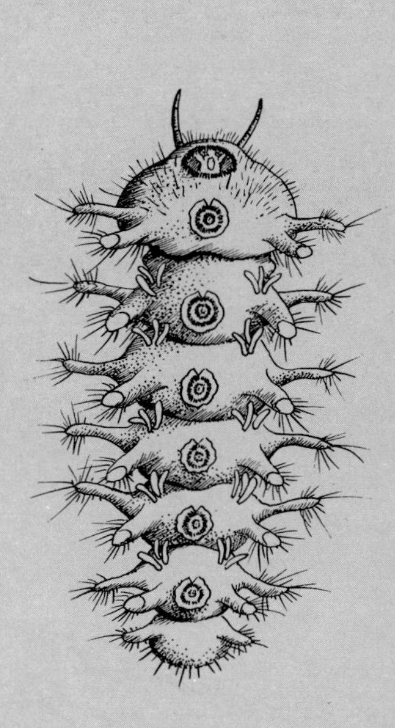

Many amphibians deposit large masses or strings of jelly-coated eggs in ponds and other still waters. The two-lined salamander, a species found in swift streams, must protect its eggs from the current; the female attaches its eggs one at a time to the sheltered underside of a rock or log in the stream.

Just as the lowly caterpillar transforms into an attractive butterfly, so also does the strange process of metamorphosis change the clumsy caddisfly larva (*above*) into a dainty winged adult like the one at the top of the opposite page.

The miracle of metamorphosis

Chances are that when you go hunting for caddisfly larvae, you will find an occasional case with the end sealed shut. Instead of the usual wriggling wormlike larva, the case will contain a strange, seemingly lifeless object that looks like a mummy. Probe it gently with a pin and the creature will squirm slightly. The insect is not dead. It is simply passing through a phase in its life history called the *pupal stage.* If you were to keep the pupa in a cool, well-aerated aquarium, the day would finally arrive when it would emerge from the security of its protective case, swim easily to the surface, and split open. A delicate mothlike creature would wriggle from the opening and fly away.

Usually dull brown or gray, with gracefully arched antennae and wings that fold tentlike over their backs, adult caddisflies are so inconspicuous that few people recognize them. During their brief existence as adults, the delicately beautiful caddisflies do not eat at all. After mating, the female returns to the water to lay its eggs. Completely surrounded by a bubble of air, it crawls into the water and deposits the eggs in closely spaced patches. More often than not, the caddisfly dies before it is able to crawl back to safety. While some species lay their eggs on submerged twigs or plant stems, others prefer to glue the eggs onto rocks, even in swift torrents where a man would have difficulty standing.

This curious process by which a creeping wingless larva transforms into a beautiful flying adult is called *metamorphosis*. The sequence from egg to larva to pupa to adult is fairly common among insects. Butterflies and moths are probably the most familiar examples of metamorphosing insects, though the larva generally is called a caterpillar and the pupa is known as a chrysalis. Beetles, flies, wasps, and several other insect groups have the same sort of life history.

A number of other insect groups are characterized by a slightly different kind of life cycle. As you search among cattails and other streamside plants, for example, you are likely to find the cast skins of dragonfly nymphs, stonefly nymphs, or other insects. These dry husks preserve in exact detail the familiar outlines of aquatic nymphs, yet the adults that emerge from them look entirely different from the immature forms.

All adult and many larval insects possess a stiff body covering. As a newly hatched dragonfly nymph begins to grow, its body eventually becomes too large to fit in its hardened skin. The nymph therefore molts. The skin splits along the back, the insect wriggles out of its old suit of armor, and a new skin hardens around its body. As it matures and grows larger, the nymph undergoes a succession of molts, and its body changes slightly in contour each time.

When the time arrives for the final molt, the nymph crawls out of the water and clings to a plant stem or other

The caddisfly's life cycle begins when an adult female deposits a patch of eggs (here magnified three times) on a stone or some other underwater site. These hatch out into larvae . . .

. . . which, as we have seen, construct elaborate protective cases and pass this period of their life cycle foraging for small plants and animals on the stream bottom. At the top of the page is a caddisfly case cut open to reveal the inch-long larva. The white structures along its sides are gills. In the lower picture, another case has been cut open to show the pupa. During this inactive, mummylike stage the larva transforms into an adult. When pupation is completed, a matter of two weeks or so, the case splits open . . .

support. The skin splits along the top of the head and down the center of the back, and the insect begins to wriggle out. With its back humped, the dragonfly squirms upward until the head and finally the whole front end are free. Then the legs and, last of all, the long, slender tail end are pulled loose. Finally, the insect spreads its transparent wings, which stiffen and dry within an hour or two of exposure to the air. As if by a miracle, the squat, ungainly nymph is transformed into a beautifully colored, fleet master of the air.

The dragonfly's life cycle, like that of damselflies, mayflies, and certain other insects, is an example of incomplete metamorphosis. These insects by-pass the inactive pupal stage and instead transform directly from the immature to the adult form. In their youthful stages, they are known as nymphs or naiads, as contrasted with the larvae of insects that undergo complete metamorphosis.

Mayflies molt twice

Mayflies have a curious extra stage found in no other insect. Once they possess full-blown wings, most insects have finished growing. They never molt again. But when a mayfly sheds its nymphal skin, its body is pale and its wings are milky rather than clear. It clings to the underside of a twig or conceals itself in some other protected spot near the water for a few hours, and then molts again. This time it emerges as a beautiful, fully developed adult.

Adult mayflies live only a few hours, or at most a few days. After months of feeding and growing in the water, the airborne adult's brief life is spent solely in mating and egg laying. Swarms of hundreds, or sometimes even thousands, gather in the late afternoon or evening over the water of streams and lakes. The swarms, mostly males, rise and fall in unison above the water, seeming to dance, with their gauzelike wings shimmering against the setting sun. Now and then a female joins the throng, mates, and lays her eggs. Some drop their eggs at random into the water, but others rise and fall rhythmically above the surface, touching the water with the tip of the abdomen and depositing a few eggs on each descent. Finally the mayflies drop exhausted to the surface and are snatched by fish as they drift listlessly downstream.

. . . liberating a caddisfly like the one below. Adult life is brief, and most species do not even eat during this stage. After mating, the female deposits eggs in the stream, and the cycle begins anew.

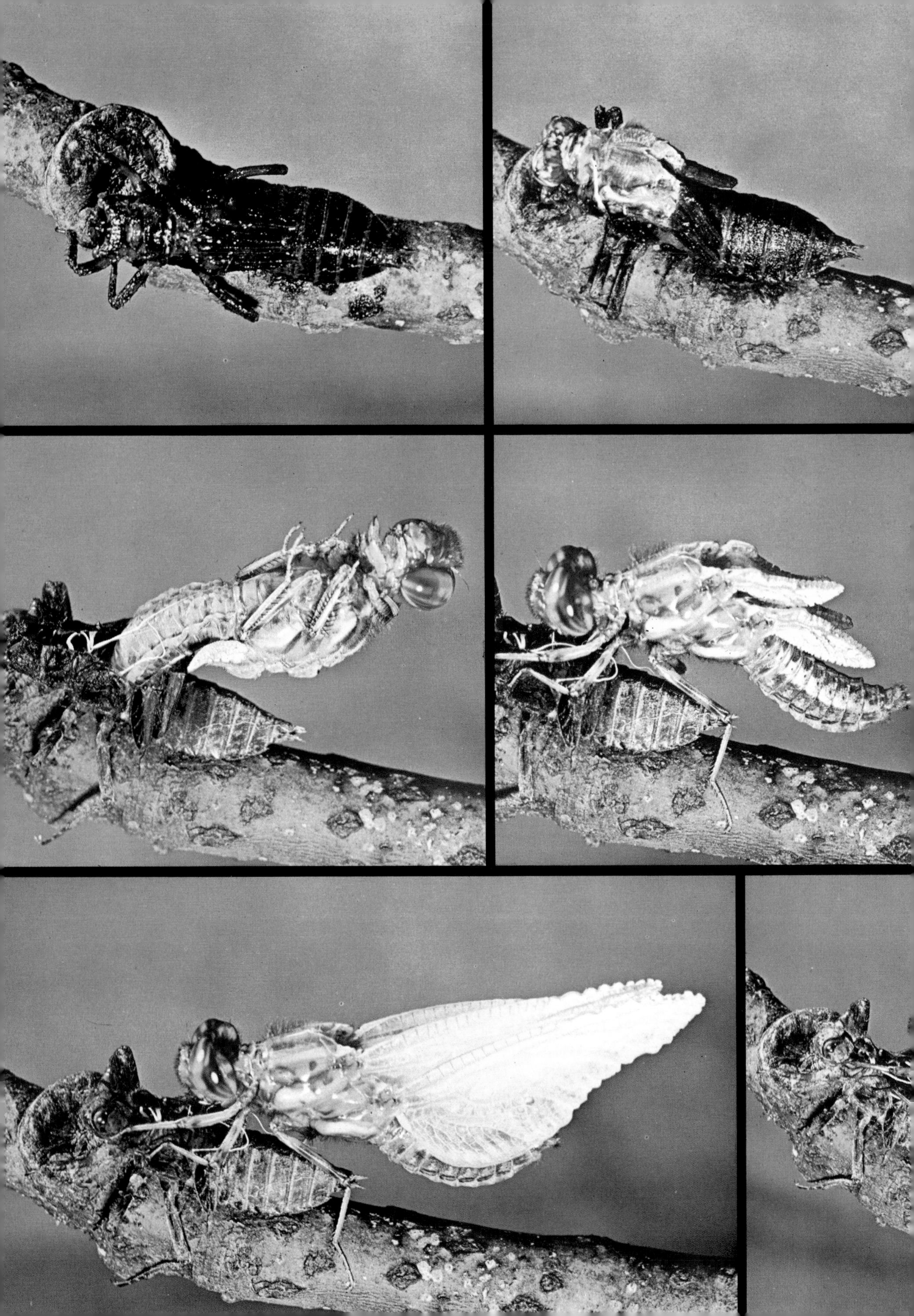

Forsaking its stream-bottom home forever, a fully grown dragonfly nymph crawls up a plant stem into the realm of fresh air and sunlight. Soon the nymph's outer skin splits down the back, permitting the adult to wriggle free. Blood surges through its rumpled wings, and before long the transparent membranes stiffen and dry in the air. Within an hour or two, a grotesque predator of the bottom mud has completed its transformation into a handsome airborne hunter *(below)*.

Mayfly nymphs, in contrast, require two molts for their transformation into adults, as pictured on the next two pages. On the first molt, a milky white subadult emerges from the nymphal skin. A few hours later, the insect sheds its body covering once again, and a beautifully patterned adult mayfly is liberated.

Life in the flowing water

The stream, as we have seen, is a composite of many habitats, from rapids and waterfalls to sluggish stretches and placid backwaters. Yet it is constant in one way: the water is always on the move. To survive in this ever-changing world, plants and animals have had to adapt their structure and habits in many ways. Some animals, such as fishes, are streamlined like airplanes. As a result, trout and other fishes are able to make their way upstream or hold their position against a strong current. Many insects, such as certain mayfly nymphs, also are streamlined.

But most aquatic animals are not streamlined at all. Rather than struggle against the force of the current, they seek to avoid it entirely. Some hide beneath rocks. Others burrow in the bottom. Many caddisfly larvae construct heavy cases, and a few even ballast their homes by cementing larger pebbles along each side. Net-winged midge larvae cling to rocks with powerful suckers, while other creatures depend on stout claws, hooks, or gluelike secretions to keep from being swept downstream.

How to get oxygen from the water, how to snatch food as it passes by in the current, how to find a mate and reproduce more of its kind: all these and other problems must be overcome if an animal is to survive in the stream. On our brief visit to one small stream, we have seen how many animals are adapted to life in this fluctuating environment. Yet so far we have concentrated on only one small part of the story. Just as you cannot know a tree by examining one short twig, you cannot know of life in flowing water by exploring only one small brook. Where does the constant flow of water come from and where is it going? What lies at the headwaters and how does a river change as it flows to the sea?

Each stream has an individuality as distinctive as that of a living thing. Each presents its own set of life-or-death challenges to the creatures that live in it. Forever changing, forever posing new problems of survival, the stream is never an easy place for life. But its community of plants and animals is always rich, varied, and endlessly fascinating.

From Source to Sea

Trace a brook to its source and very likely you will come eventually to a spring. At the base of a rocky cliff, perhaps, or in a thicket of willows or alder, you find a pool of cold, sparkling water that bubbles up from the earth. A catbird chatters noisily from the thicket, and in the distance a vireo is singing. Then all is silent. A thread of water flows unendingly across the lip of the pool, but makes barely a sound as it trickles down the gentle slope. Only when the stream has joined forces with others and has begun to tumble across steeper slopes will its whisper become a full-fledged song.

Because of perpetual dampness, the spring is surrounded by lush greenery. Thick mats of moss cushion the surface of every rock. If it is April or early May, a fringe of marsh marigolds may encircle the pool with splashes of brilliant yellow, or clumps of ferns may form a luxuriant border of green. Fine strands of algae, like tangled mats of green hair, sometimes all but clog the spring, and masses of water cress may poke their tender leaves from the shallows.

If the bottom of the pool is covered by silt, you probably can see where the water bubbles in; flecks of sediment glint in the sunlight as they bob up and down on currents of

water that well up from the bottom. Dip your hand in the water and you find that it feels cool, at least in summer; in winter the water feels warmer than the air. Since the water issues from deep in the earth, its temperature changes little with the seasons. As a result, plants growing in the spring and the upper reaches of the stream may be green all year round. On dry hillsides or desert slopes, springs are surrounded by green oases that may be visible for miles.

In August and early September brilliant spikes of cardinal flower are a common sight near spring-fed streams. Hummingbirds often visit the plants and probe for nectar hidden deep in the flowers' tubular throats.

Life in a spring

The relatively moderate, constant temperature and dependable flow of water would seem to assure abundant life in the spring. True, there are plenty of visitors to the spring: grackles and jays, deer mice, skunks, and other creatures that come to sip its water. But animal life in the pool itself is rather sparse, mainly because the water is relatively poor in dissolved oxygen.

The first creatures to catch your eye may be water mites, tiny relatives of spiders. Their plump oval bodies, seldom much larger than pinheads, usually are bright red or orange. Water mites spend much of their time creeping on the bottom or clinging to plant stems, but they can also swim by vigorously flailing their eight hairy legs. Despite their small size, most water mites are active predators that live by sucking juices from small prey.

If you stir the plants, the spring is likely to swarm suddenly with another form of life. Throngs of pale shrimplike creatures about half an inch long seem to appear from nowhere. Known as scuds or sideswimmers, they dart to and fro on their sides by rowing rhythmically with their tiny legs. Scuds usually hide by day in tangled vegetation and come out at night to feed on dead plants and animals. So prolific that a small spring may harbor thousands of them, scuds serve as food for all sorts of larger animals.

Besides snails, flatworms, and a few beetles and other insects, the spring may also contain salamanders. One of the commonest is the spring salamander, which is pink,

The glossy butter-yellow blossoms of marsh marigold, or cowslip, brighten the edges of springs and spring-fed streams in late April and early May, usually before the leaves are out on trees.

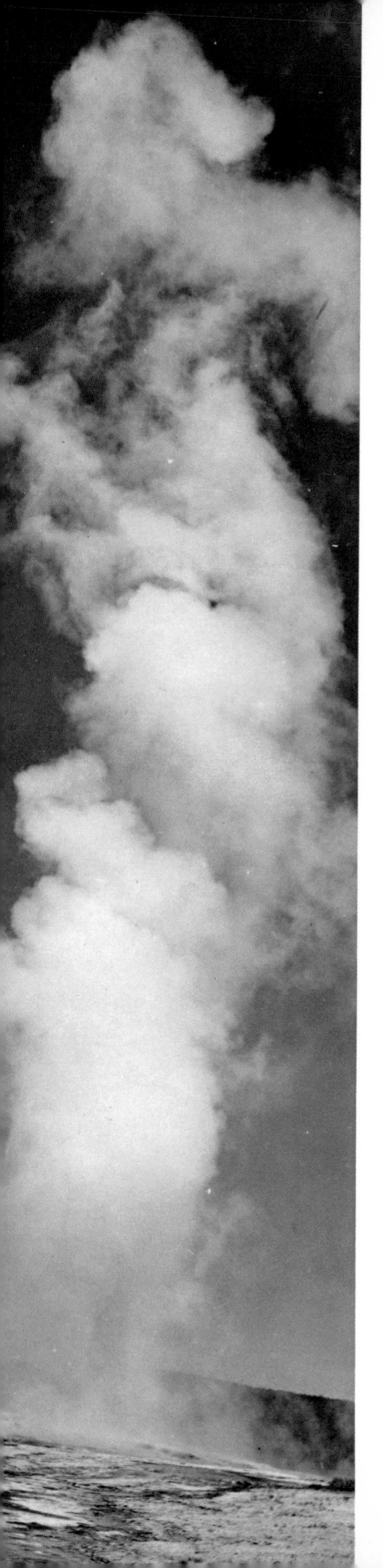

salmon, or brownish in color. Living in cold streams as well as in springs, it lurks under stones and in crevices during the day and emerges after dark to hunt for insects and other small animals.

Hot springs and geysers

A very special habitat is provided by the springs in Yellowstone National Park, Hot Springs National Park in Arkansas, Death Valley, and several other areas throughout the country. Instead of the cool, refreshing water we usually associate with mountain springs, water that pours from the earth in these places is steaming hot. At Hot Springs, Arkansas, for example, forty-seven springs provide a daily flow of nearly a million gallons of water with a constant temperature of 145 degrees Fahrenheit. Water in some of the springs at Yellowstone is 194 degrees; at the 7000- to 8000-foot elevation where these springs are found, this is almost the boiling point!

No one knows exactly why hot springs occur or how they work. But we do know that in areas of volcanic and hot-spring activity the earth's inner heat comes closer to the surface than elsewhere. In several places at Yellowstone, for example, the ground is warm or even hot to the touch. When scientists drilled beneath the surface at Yellowstone's Norris Geyser Basin, they found a temperature of over 400 degrees just 365 feet below the surface. As underground water seeps through these rocks, it becomes very hot.

In some places the superheated water is under such extreme pressure that the steam seeks a way out. Sometimes the steam and other gases escape in steady jets through vents known as *fumaroles.* Hot springs from which water escapes in periodic bursts are called *geysers.* "Old Faithful," the most famous geyser, usually erupts every sixty-one to sixty-seven minutes and expels about ten thousand gallons of water before it subsides four or five minutes later. As if by clockwork, a low warning rumble is followed by a few short spurts. Then, as the eruption reaches its full force, a column of water rises 120 to 170 feet, topped by clouds of steam that form as hot vapor strikes the cool air.

Yellowstone's "Old Faithful," the world's most famous geyser, hurls ten thousand gallons of scalding water high into the air at approximately hourly intervals.

Dissolved limestone in the near-boiling water of Yellowstone's Mammoth Hot Springs is deposited in massive steplike formations of travertine, a mineral very similar to those that form the stalactites and stalagmites found in caves. Heat-tolerant algae tint the stony surfaces with delicate pastel hues.

In places where the steam escapes through soft earth, bubbling "mud pots" form. As bubble after bubble sputters and bursts through the surface, the whole mass resembles a great pot of pudding cooking on a stove. In other places steaming pools are enclosed by steplike tiers of stony sculptured walls. These so-called travertine terraces are deposits of minerals that were dissolved in the water.

But what about life? Can plants or animals survive the near-boiling temperatures in hot springs? At Yellowstone's Mammoth Hot Springs, where you will find one of the most beautiful of the travertine terrace formations, water flows over red, pink, tan, brown, and bluish gray rock surfaces. If you were to examine scrapings from the rocks under a microscope, you would see that the color results from the presence of thousands of minute algae that flourish at a temperature of 170 degrees.

Life in hot springs

Of all living things, algae can withstand the highest temperatures. In certain hot springs at Yellowstone, tiny one-celled blue-green algae thrive in water at 194 degrees.

YELLOWSTONE NATIONAL PARK

Although Yellowstone is best known for its geysers and hot springs, thermal phenomena are not the only attractions at this magnificent national park. Situated a mile and a half above sea level, the park includes 3472 square miles of spectacular Rocky Mountain wilderness. High on the list of scenic wonders are the Lower Falls of the Yellowstone River (*right*), where the water plunges over a 308-foot-high precipice before racing down the 28-mile-long, 1200-foot-deep Grand Canyon of the Yellowstone. In the heart of the park lies the largest mountain lake on the continent. Over a thousand miles of hiking and riding trails thread through unspoiled wilderness. The park's birds —240 species—range from the majestic bald eagle to the comical little dipper. Among the park's mammals are North America's finest: American elk, with gracefully branching antlers (*upper left*); moose, with massive flattened antlers (*lower left*); bison; mountain sheep; and of course bears, which are as much a symbol of the park as Old Faithful itself. Yellowstone is the oldest of our national parks, and in the eyes of many visitors it is also our finest.

Animals are much less tolerant of high temperatures. Even so, certain one-celled animals, or protozoans, live where the water has cooled to the still quite warm temperature of 130 degrees. Water mites have been found where the water is 122 degrees, and certain midge larvae can withstand a temperature of 120 degrees.

As streams flow away from hot springs and geysers, the water cools rapidly. Within a few feet of their sources, they begin to support more midge larvae plus a variety of mayfly and dragonfly nymphs, caddisfly larvae, water beetles, creeping water bugs, and other common insects. Even so, life is relatively sparse. Fish, for example, are not found in any of the very hot springs at Yellowstone. A few small fish do

A stream of hot water tinted with heat-tolerant algae issues from one of Yellowstone's innumerable hot springs and meanders across a plain of mineral-encrusted mud. Parts of this crust are sturdy enough to walk on, but thin spots sometimes give way suddenly beneath the weight of a man or animal, plunging the trespasser into a pool of hot water.

occur in certain warm springs in the Death Valley area, but they are relicts of a time when warm waters were distributed far more widely over the western parts of North America.

The scarcity of animals in this seemingly favorable habitat is easily explained. Besides having an abnormally high mineral content, the warm water contains relatively little oxygen, for the hotter water becomes, the less dissolved oxygen it can hold. Yet most living things need even more oxygen at high temperatures than at low ones, since the rate of their body processes increases at higher temperatures. Thus, in warm water, where oxygen is most scarce, the oxygen requirements of animals usually are greatest.

The water cycle

Where does the water in a spring come from? Certainly it does not originate beneath the surface of the earth. Like most of the water in the world—in the sea, lakes and ponds, swamps, rivers, and streams—it is part of an enormous, endless cycle, so vast as to stagger the imagination.

The great reservoir for most of the water on earth is the ocean. There, with energy supplied by the sun in the form of heat, water evaporates and enters the atmosphere. As a vapor, it rises with warm air until the air begins to expand and cool as it reaches higher altitudes. Since cold air cannot hold as much moisture as warm air, the water vapor condenses into tiny drops of liquid that are visible as fog or

A veil of morning mist hanging over a placid river is striking evidence of water's unending cycle. Evaporated from the surface of the river, the moisture has condensed into tiny droplets that are visible as fog, which is nothing more than a low-lying cloud.

clouds. As the process continues, the drops grow larger and finally fall as rain. Under certain conditions, the moisture may form snow, sleet, or hail. The water may return directly to the ocean; or, if the clouds are carried over continents by the winds, it may fall on land, only to be carried back to the sea in rivers and streams. There, as the water evaporates, the cycle begins once again.

Water always returns eventually to the sea, though it may be detoured at any step along the way. It may be stored for months or even years in the form of ice in glaciers, or its journey to the sea may be interrupted by a lake. Before it reaches the sea, the water may evaporate from lakes or streams and then fall elsewhere as rain. A great deal of water also is stored in the bodies of living things.

THE WATER CYCLE

Constantly in motion as it makes its way toward the level of the sea, the water in rivers and streams, like every drop of moisture on the earth, is part of a vast eternal water cycle. Every year, in an expenditure of energy beyond the mind's ability to comprehend, heat from the sun lifts eighty thousand cubic miles of water from the sea in the form of water vapor. Water also is evaporated from the surfaces of lakes and rivers, while still more vapor is given off to the atmosphere by the breathing of living animals. Finally, through the process of transpiration, plants give off staggering quantities of water vapor.

The moisture then condenses and forms clouds, many of which are carried over land masses by the wind. There the water falls as rain, snow, or other forms of precipitation and flows once again toward the sea in rivers and streams. Or the moisture may seep through the soil and be stored temporarily as underground water. Even this vast reservoir is not lost permanently to the water cycle, however. Unable to pass through impervious rock layers, it may reemerge at the surface in the form of springs, or it may flow out into marshes or directly into stream channels.

SUN
CLOUDS
TRANSPIRATION
EVAPORATION
OCEAN
SURFACE RUN-OFF
EVAPORATION
LAKE
SPRING
IMPERVIOUS ROCK LAYER
MARSH
UNDERGROUND WATER

Instead of running off the surface in rivers and streams, however, much of the water seeps into the earth, where it fills the space between soil particles. Even so, some of this water returns almost immediately to the atmosphere. It is taken up by the roots of living plants and then, by the process called *transpiration*, is given off to the air in the form of vapor that escapes through pores in their leaves.

But most of the underground water continues to seep down through the soil. Near the surface many spaces between soil particles are filled with air. As the water continues to trickle down, it eventually reaches a zone of saturation, where every space in the soil and every crack or crevice in the underlying rock is filled with water. This is the water that is tapped when wells are drilled. The top of this saturated zone is commonly called the *water table.* It may lie at or near the surface, or it may be as much as two miles down. In most places, however, there is little water beyond a depth of half a mile.

At Big Spring State Park in the Missouri Ozarks, a quarter of a million gallons of water gush up from the earth each day, providing dramatic evidence of the large amounts of water stored in underground rock layers.

The total amount of water stored underground is enormous. Scientists estimate that throughout the United States, the total usable underground water supply is equal to about ten years' annual precipitation. Yet even underground water is not lost permanently from the overall water cycle. Beneath the porous rocks that permit seepage of water lie other impervious layers that act, in a sense, like dams. When underground water reaches them, it can no longer move directly downward. But as surface water continues to filter in from above, the underground water begins to flow sideways along the sloping contours of the impervious rock layers. At lower elevations, the underground water eventually emerges on the surface. Sometimes it flows directly into channels of rivers and streams, maintaining their flow between rainy periods and snowmelt seasons. Or it may ooze out in the form of springs.

Not all springs are like the one we visited. Sometimes the water escapes through crevices high on the walls of cliffs and trickles down their faces. Seepage from springs of this sort at Zion National Park in Utah results in beautiful "hanging gardens." The plentiful water supply supports lush growths of ferns and wild flowers—columbine, shooting star, scarlet lobelia, and many others—in an otherwise arid region. Other springs, such as those in areas of Florida, are as big as small lakes. Their overflow—sometimes millions of gallons a day—runs off as full-fledged rivers right at the source. Or underground water may simply ooze to the surface over broad marshy areas.

Rivers of darkness

In some places underground water flows as rivers. In the southeastern United States, whole cave systems extending for hundreds of miles have been dissolved out of limestone by the action of underground water. One of the most famous is at Mammoth Cave National Park in Kentucky, where visitors can take boat rides on Echo River, 360 feet beneath the surface of the earth. But even rivers that flow in caves ultimately emerge on the surface and find their way to the sea.

Although they are shut off from the energy of the sun and therefore lack the essential productivity of green plants, cave rivers support a sparse and curious community of animals: colorless fish and crayfish, ghostly pale salamanders, flatworms, and a number of others. Living in perpetual darkness, where eyes are useless, most species have become blind. Color likewise is of no advantage to animals whose mates or predators cannot see, and most of them are colorless. Yet other sense organs have become so highly developed that they compensate for loss of vision. The heads and sides of blind cave fish are studded with supersensitive protuberances that detect vibrations in the water, thereby

This blind, nearly colorless salamander is a member of a bizarre community of creatures that live and die in the endless night of underground rivers. Eyes and color patterns have no value here; natural selection, operating over thousands of generations, has eliminated them.

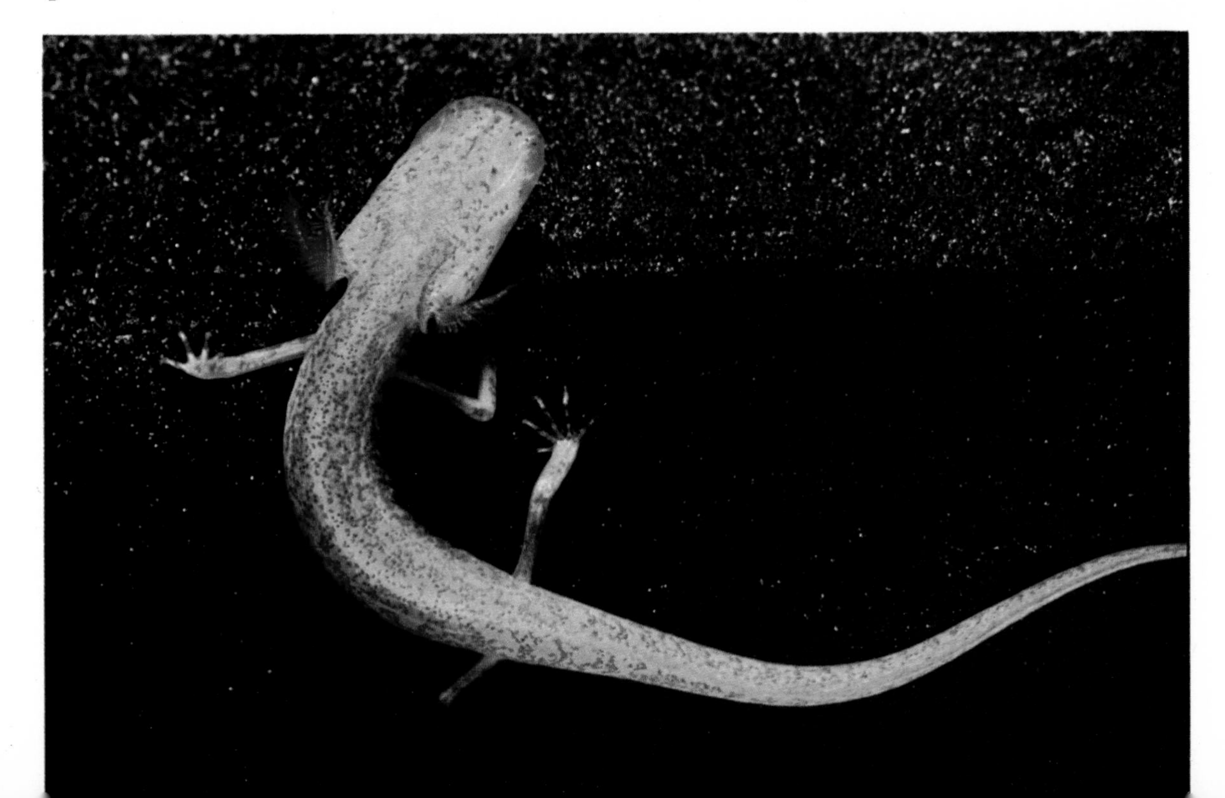

enabling the fish to locate prey. Cave crayfish grope through darkness by tapping the stream bed with extra-long antennae that feel and possibly taste and smell the surroundings. Never abundant, almost all cave dwellers lead sluggish lives and exist on sparse food supplies that are washed in from the surface.

In general, a great deal is known about the life in cave rivers that are readily accessible. But so far we have had only tantalizing glimpses of the life in other underground water stores. Occasionally a blind salamander or fish, for example, creates a sensation when it is pumped up with the water from a deep well.

Snowpack and surface water

Although underground water is probably the greatest store of fresh water in the world, many rivers are fed largely by surface water. In the North and in the eastern half of the United States, rain usually falls every month of the year. Where the land is flat, the soil becomes so saturated that the water table lies literally at the surface. Lakes and swamps are everywhere, connected by networks of rivers that form extensive inland waterways.

Superior National Forest on the Canadian border in northern Minnesota is such an area. In this land that has been

The weird cry of the common loon, drifting like a lunatic's laughter across a lake at midnight, has sent a shiver down many a camper's spine. Light of day reveals an attractive water bird dressed in handsome black and white. In winter, both male and female exchange their checkerboard breeding plumage for sedate grays and white.

In his iridescent breeding finery, the male wood duck is one of the most handsome of all American waterfowl. At one time the bird was nearly exterminated by excessive hunting and the destruction of many nesting sites, but in recent years conservation practices have led to a dramatic comeback in wood duck numbers.

scoured and grooved by ancient glaciers, islands of fragrant evergreen forests dot a region that seems to include more water than land. Placid lakes and shimmering ponds are hemmed in by sheer granite cliffs and connected by a maze of meandering streams. A canoeist could travel for days—perhaps weeks—and never need to camp twice on the shores of the same lake.

On a trip through this watery wilderness, you might see moose plunging their heads in the shallows for lily roots and other succulent plants. Minks, otters, raccoons, and a host of other animals thrive in this land of trees and trout-filled water. Ducks might precede your canoe around each bend, and in autumn great wedges of geese would pass overhead, winging their way from breeding grounds in the Arctic to winter homes on the Gulf Coast. A hole in a tree beside the stream, perhaps the work of a pileated woodpecker, might be occupied by a delicately beautiful wood duck brooding a clutch of a dozen or so eggs. With luck, you might even see the newly hatched ducklings leaving the nest. Soon after hatching, the ducklings struggle one by one over the lip of the hole and drop to the ground, perhaps forty feet below. Unharmed, they scramble to the water and swim away.

A snow survey worker takes a core sample with a hollow tube. By measuring the depth of the snow and then weighing his samples, the scientist gathers information that is helpful in predicting the spring run-off of meltwater from mountain snowfields.

In other parts of the country a great deal of the annual precipitation falls as snow, especially on mountains in the Far West. Moisture-laden winds blow in from the Pacific and drop record amounts of snow on the Olympic Peninsula in Washington and on the Sierra Nevada in California. As much as 73½ feet of snow have fallen on parts of the Sierra Nevada in a single season. This is equal to a great deal of rainfall, since it takes only a six-inch layer of moist snow or a thirty-inch layer of dry powdery snow to equal the water in a one-inch rainfall.

As the snow accumulates, it forms a pack. During warm spells, as some of the snow melts, the water seeps down and refreezes into hard granular masses of ice crystals. In spring and summer, the pack melts gradually and keeps mountain streams brimming with cold water well into summer.

In the West, where water is scarce, this supply is so important that scientists conduct snow surveys in the mountains each winter. By measuring the snowpack, they can predict the amount of run-off that will be available for irrigation in the summer. Occasionally, however, the snowpack does not release its water gradually. A sudden warm spell early in spring may cause most of the snow to melt at once. Disaster follows as rivers swell and flood beyond their borders, leaving a wake of destruction all the way to the sea.

Rivers of ice

A visit to Glacier National Park in Montana or to Mount Rainier National Park in Washington provides an opportunity to see streams and small rivers that spring from a sort of permanent snowpack. Here and in a few other places, shrunken glaciers still exist high in the mountains. They are frail remnants compared with the colossal sheets of ice that crept slowly across Canada and much of the northern United States during the last ice age, a mere ten or twelve thousand years ago. But they are genuine glaciers, rivers of ice and stone that slip slowly down the mountainsides as more snow accumulates each winter.

Among the easiest to visit are those in Mount Rainier National Park, where twenty-six glaciers cover more than forty square miles. It is only a short walk from Paradise Valley, for example, to Nisqually Glacier. As you walk up

Nisqually Glacier rests its dirt-covered snout in the valley it has carved from the flanks of Mount Rainier. Heaps of rubble deposited farther downslope reveal that Nisqually is a receding glacier, one whose forward movement cannot keep pace with its rate of melting.

Two climbers make their way across Winthrop Glacier, one of twenty-six rivers of ice in Mount Rainier National Park. The enormous cracks, or crevasses, develop as the glacier slips ponderously over the brink of an abruptly steepened slope. Bridged over with a light covering of snow, a hidden crevasse can become a deathtrap for an unwary mountaineer.

the trail, you may not recognize the lower end of the glacier. It is simply a tremendous heap of dirt and stones mixed with ice. But this is typical of shrinking glaciers, for they can be thought of as rivers of rock as well as of ice. Plucked from the mountainside by the mass of moving ice, the rocks acted as the glacier's teeth, scouring the surface and carving the landscape. Now a warming climate is causing Nisqually Glacier to retreat as much as seventy feet a year. As the ice melts, the boulders are stranded in jumbled heaps along the glacier's snout. Farther up the mountain, the glacier is much thicker and looks much cleaner.

At the foot of the glacier, a stream gushes from the boulders and debris. This is the beginning of the Nisqually River, and there can be no doubt of its origin; its source is meltwater from the glacier itself. The water is gray and muddy with rock powder, bits of stone ground fine as flour by the moving ice sheet. Farther downstream the powder settles out and the icy water becomes crystal clear.

Each afternoon in summer, the water reaches a flood stage because of the tremendous amount of melting, and then it subsides overnight. Because the water is cold, it can hold a great deal of oxygen, which is ideal for living things. On the other hand, the water is poor in food materials since it has just poured from a glacier. If you pick up a rock from the stream, you will find an assortment of stoneflies, mayflies, and other typical stream animals. But they will be less abundant than in other mountain streams, since the rock has only a sparse coating of microscopic algae to serve as food.

Strangely enough, a few plants and animals survive even on the ice of the glacier itself. Besides a few specialized algae, much of the surface is dusted with wind-borne pollen grains that have blown up from evergreen trees on the lower slopes. The pollen provides food for throngs of glacier fleas, a kind of springtail. In summer, thousands of the tiny insects hop around in great swarms. The only other animals on the glacier are a few harvestmen, or daddylonglegs; now and then you may see one scurry across the ice in search of springtails. Thus, even in this frozen world, the lives of all the inhabitants are linked in a chain of interdependence: thousands of springtails eat the pollen, and a much smaller number of harvestmen eat the springtails.

A stream of meltwater tunneling beneath a glacier was the major force in carving this spectacular ice cave on Mount Rainier. On summer days, glacial streams like this one reach their peak flow late in the afternoon and then subside as the sun sets.

River patterns on the land

Few streams, other than very short ones along the coast, flow directly into the sea. Most rise from sources that lie hundreds of miles inland. As they journey to the sea, small streams flow into larger ones, and the large streams then unite to form rivers before the water finally finds its way back to the ocean. Trace any river system on a map and you will see that it assumes the general pattern of a tree. The main river forms the trunk, smaller tributaries resemble branches, and a maze of brooks and streams fan out like twigs in the headwater areas.

As any stream or river is joined by tributaries draining other areas, the volume of water it carries grows steadily larger. To accommodate this additional flow of water, the river's channel gradually increases in width and depth. Strangely enough, actual measurements have shown that the velocity of the current remains very nearly constant throughout the river system, even though mountain creeks and rapids may appear to flow more swiftly than the deep water of large rivers.

As they journey from their headwaters to the sea, moreover, streams and rivers sometimes are described as being young, middle-aged, or old. These terms refer not to time but to the condition of the landscape carved by the river. In its youth or at the headwaters, a river is eroding downward so rapidly that it flows through deep, steep-sided, V-shaped valleys. As it matures, it flows with gentle curves through broad valleys bordered by smoothly rounded hills. In old age, the river flows in broadly meandering curves across a nearly flat plain; the surrounding mountains and hills have been almost completely worn away. Here, especially, the river frequently spills over its banks and leaves rich deposits of silt on the flood plain.

The progression from headwaters to sea, of course, is not invariable. Small upland streams may meander across flat meadows just as great rivers usually snake across level plains near the sea. But the sequence generally holds true.

A photograph taken from the *Gemini 4* spacecraft orbiting over western Texas shows with remarkable clarity the characteristic treelike pattern assumed by river drainage systems. The faintly visible rectangles are cultivated fields on farms and ranches.

Downstream

The best way to understand the majestic flow of water from mountain to sea is to observe it firsthand, perhaps by traveling in a canoe or rubber raft. In most rivers you would not have to paddle at all, for the force of the current would carry you effortlessly downstream. Few people have the opportunity to follow a river through its entire course. But even a few hours spent drifting lazily down a medium-sized river can be a rewarding experience.

One of the first things you might notice is the color of the water. Instead of having the clear sparkle of a mountain stream, the water probably is cloudy or even muddy. If you were to leave a jar of river water standing on a shelf, it would clear eventually. But the bottom of the jar would be covered by a layer of fine silt. Besides making river water less attractive in appearance than the clear water of a mountain stream, the load of silt, as we shall see, has profound effects on life in the river.

Where does all the silt come from? A great deal of it is carried into the river by tributaries. But some of it is eroded from the river's own banks. As you round a bend, your canoe probably is swept along the outside of the curve, for here the water flows fastest and does its greatest work of removing soil. You notice that the bank is steeper and the water much deeper on the outside of the bend. Undercutting of the bank may even have caused a great tree—very likely a willow, a cottonwood, a sycamore, or even a pine tree—to topple into the water.

For the kayak enthusiast, the white water and rapids on a mountain river are an invitation to adventure. Other people prefer slower rivers where they can explore at leisure as they journey downstream in a canoe or rubber raft.

Much of the soil that is eroded from the banks will not be carried all the way to the sea, however. Just as silt settles from the calm water in a jar, part of the river's load of mud and sand is deposited on the bottom wherever the current slows—when the river enters a broad lakelike stretch, for example, or in the shallows on the inside of a curve, where sand bars form. In this way the river constantly changes course as curves shift from place to place with the passage of time.

Otter antics

No matter where you travel, the riverbanks are likely to be fringed with various kinds of trees. In the Great Plains, rivers are usually bordered by walls of great cottonwoods and willows, while, just beyond, the prairie rolls on for miles with not a tree in sight. In humid areas, the stretches between farmlands, villages, and cities are nearly always occupied by forests—evergreens in the West, the Far North, and parts of the South; mixed hardwood forests in most other areas.

River otters occur over most of North America, but they are rare and wary of contact with humans. Even in areas where otters are relatively abundant, outdoorsmen are far more likely to encounter their characteristic tracks than to see the animals themselves.

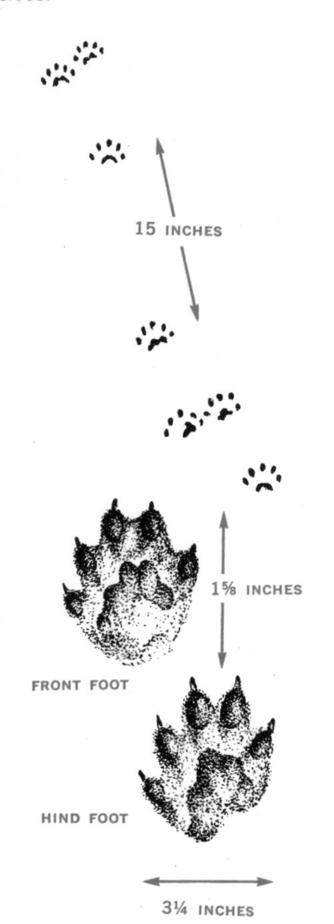

As you drift quietly through a wild, forested area, you may be fortunate enough to glimpse an animal swimming swiftly against the current. Only its head and part of its back project above the water, and a great V-shaped trail of ripples streams backwards from its chin. Sensing your presence, the animal slips suddenly beneath the surface, leaving only a slight eddy to mark the spot where it disappeared. And that is the last you see of it.

The animal, about four feet long, was a river otter. Although otters are found in nearly every state and sometimes live near populated areas, they are seldom seen. These large relatives of minks and weasels have been so heavily trapped for the sake of their rich, lustrous fur that they are no longer common anywhere. The few that remain are secretive in their habits, venturing from their dens mainly at night.

Beautifully adapted to life along rivers, streams, and even lakes, the otter has oily fur that forms a sleek waterproof coat. Webbing between the toes transforms its feet into efficient paddles, while the long, tapering tail serves as a rudder. Sinuous undulations of its streamlined body add to the otter's speed and agility as a swimmer. When it dives, special flaps make its ears and nostrils watertight. And dive it does; the otter can remain submerged four minutes or so without coming up for air. In winter it even swims under the ice, surfacing occasionally to tap the pockets of air trapped between the water and the ice.

The otter eats frogs, crayfish, a duckling now and then, and even large insects, but its favorite food is fish. This

Enjoying the best of two worlds, a sleek, slim river otter emerges from the water for a breath of air. Although otters are superbly adapted to life in water, they are very much at home on land as well.

THE CLOWNS OF THE RIVERS

River otters seem to be endowed with an abundance of good humor and joy of living. Porpoising through the water in a game of follow-the-leader, tumbling and wrestling together in the grass along the bank, tobogganing down a snowslide or mudslide for hours on end—much of the otters' daily activity seems to be just for the fun of it.

Short legs and a long body make the otter somewhat ungainly on land, but in the water these same characteristics make the animal a swimmer of unmatched speed, grace, and ability. As lithe as an eel, the otter usually gets what it goes after—most frequently a fish, although frogs, crayfish, birds, and small mammals are also welcome prey. The otter shown here is hot on the trail of a bass.

Winter weather does not inhibit the otter in the least. If the river is frozen over near its den, the otter scoots and slides through the snow to the open water of a riffle or waterfall and plunges in, oblivious to the bitter cold. The secret to its hardiness lies in its wonderfully thick water-repellent fur.

The female otter bears a litter of from one to five young in early spring. For the first three months of their lives, the pups are so jealously guarded by the mother that not even their father is allowed to approach his offspring. Later, both parents take a hand in raising the young and teaching them the basics of otter life: swimming, hunting, and, of course, playing games.

At the first hint of danger, a wary muskrat tumbles from its waterside perch . . .

agile swimmer can outmaneuver even trout, although other fish, such as crappies and suckers, are just as important in its diet. With its victim clamped between its jaws, the otter surfaces and retreats to a sheltered spot along the bank to eat. Curiously, the otter almost always eats fish from the head back, swallowing bones and all, but discarding the tail.

Even if you never see this talented fisherman in action, you may discover telltale evidence of its presence in many areas. Besides tracks and partly eaten remains of fish, the playful otter sometimes marks the riverbank with slides. Where the river is bordered by steep banks of mud or clay, groups of otters often clear away sticks and other debris to form a smooth chute. Then with forepaws tucked back and hind legs trailing, they sprawl on their bellies and toboggan to the bottom. They return again and again to the top, seeming never to tire of this sport. In winter they construct their slides in snowbanks, often with an icy pool of water at the bottom.

Of muskrats and mussels

Although otters are a rarity in most areas, muskrats are fairly common along many slow-moving rivers. Plump as overgrown meadow mice—the two, indeed, are closely related—muskrats are nearly as prolific. In the course of a summer, a single female may bear as many as three litters of five to ten young at a time. Thus in spite of heavy trapping by man and predation by minks and owls, muskrats continue to flourish throughout the country.

In shallow marshes, the muskrat's domelike houses built of cattails and other plants are a familiar sight, but along rivers its dens are more difficult to find. Instead of building mound-type houses, which the current would too easily damage, the versatile muskrat burrows into riverbanks, digging like a terrier with its forepaws and shoving the dirt out with its hind feet. The den entrance is well camouflaged since it usually opens below water level. Beyond the entrance, the tunnel slopes upward to a snug dry chamber, which the muskrat lines with plants. Another tunnel leading to the top of the bank provides both ventilation and an escape hatch in case the den should be raided by a hungry mink.

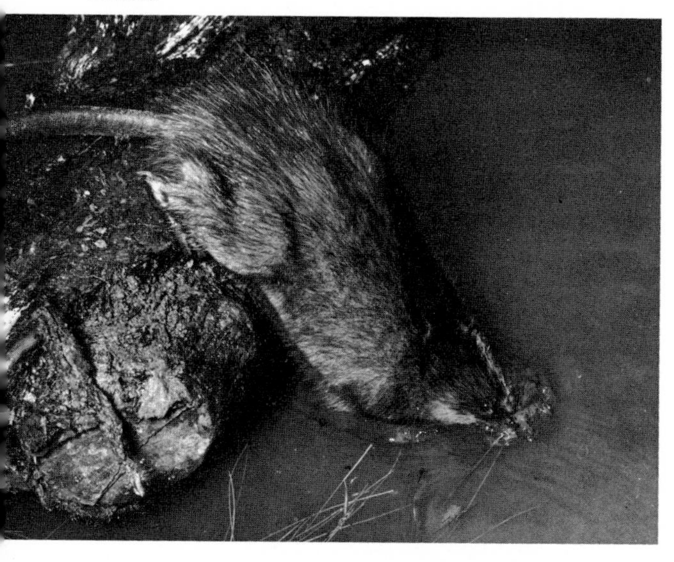

. . . and, making hardly a ripple, heads for the safety of its den—in rivers, a burrow in the bank; in standing water, a hollow mound of plant material plastered together with mud.

Like the otter, the muskrat is an expert swimmer. A trace of webbing and fringes of stiff bristles on its toes increase the surface area of its paddlelike hind feet. Its scaly ten-inch tail is flattened from side to side and seems to be useful for steering. As it patrols upstream and down in search of succulent plants to eat, the muskrat usually swims with just its nose and the top of its head projecting from the water.

The muskrat varies its diet from time to time by diving to the bottom to hunt for mussels. It sometimes uses its strong front teeth to force open the mussel's tightly closed shell. More often it simply drops the shells on dry land; when the mussel dies, the two valves of the shell open by themselves and expose the morsel of flesh inside. The muskrat is so fond of mussels, in fact, that sizable heaps of empty shells frequently accumulate at its feeding stations.

Mussel habits

If you find a heap of discarded mussel shells along the riverbank, take a closer look at them. Like clam shells, they are hinged along one side. Each half closes perfectly against the other, forming a tight protective case. The concentric ridges on the outside of the shell mark periods when the mussel's growth was slowed by winter, drought, or other unfavorable conditions. Inside, the shell is smooth and pearly.

The mussel filters food from the water pumped constantly through its body chamber. Mucus on its gills traps food particles, while constant beating of minute hairs on the gills moves the food forward to its mouth. By some process not fully understood, the mussel can detect and reject inedible materials.

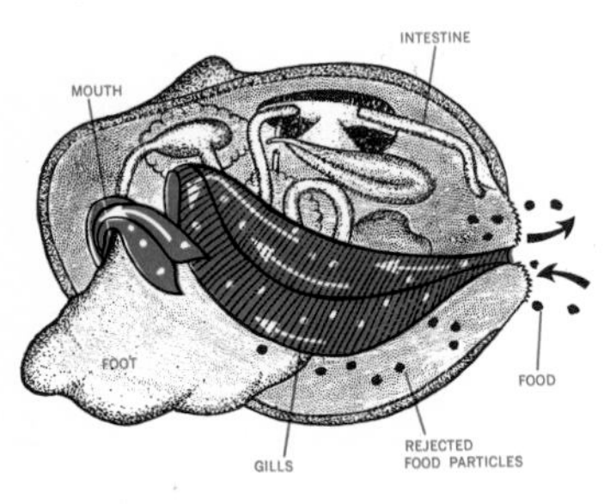

At first glance the soft body enclosed in the shell of a living mussel seems a formless mass of flesh. Most conspicuous is the foot, a muscular protuberance near the center. The mussel extends the foot by pumping blood into it, and then contracts the muscles to pull itself forward. But most of the time the mussel simply sits on the bottom, hinge end up, with its shell partially buried in fine pebbles or sand.

When it comes to feeding, the mussel takes the easy way out and lets the water bring its food. With its shell opened slightly, the mussel pumps a steady stream of water through its body chamber. The water enters through one slitlike opening, or siphon, at its rear and leaves by another just above. A coating of mucus on its gills traps microscopic plants, animals, and bits of decaying material carried in with the water. Constant beating by minute whiplike hairs on the gills keeps the ribbon of mucus and food moving forward to the mussel's mouth, while wastes are pumped out with the departing current of water.

How mussels multiply

Water also is important to the river mussel for reproduction and, strangely enough, so are fish. Since mussels usually live in dense beds on the bottom, fertilization of their eggs is no problem. Males simply release clouds of sperm into the water, depending on currents to carry it to nearby females. But once the eggs are fertilized, strange things begin to happen.

Fertilized eggs—hundreds of thousands or even two or three million at a time—are released into chambers in the female's pouchlike gills. After a short period of development, the eggs hatch and larval mussels by the thousands pour out into the water and settle to the bottom. These curious creatures, so tiny that they are scarcely visible, bear only a slight resemblance to the adult. Although the body is enclosed in a hinged shell-like covering, the free ends of the valves are armed with sharp hooks, and a long, slender thread dangles from the body.

Many of the larvae die and others are devoured by larger animals, but a good many survive to complete their life cycle. The secret to success lies in the constant snapping of the shells. If a fish happens to stir the bottom sediments, the larvae clamp their hooked shells tightly to its fins, body, or gills. The thread hanging from each mussel's body may also help it to attach itself. Before long, the larvae are embedded in the flesh of the fish, where they live as parasites, taking nourishment from the host's body fluids. Although a single fish may carry dozens of larvae at a time, it does not seem to be seriously harmed by the parasites. After several weeks of growth, the larvae gradually mature, burst from their protective cysts, and drop to the bottom as miniature adults.

Curiously, the larvae apparently are able to determine the kind of fish to which they have attached, and various species seem to parasitize only certain kinds of fish. If they have hitched a ride on the wrong species, they soon drop to the bottom and wait for another host to come within reach.

This complex life history serves a useful purpose. If the larvae all grew where the mother expelled them into the water, they soon would crowd each other out of existence. And if they depended on the current to distribute them over the bottom, many would be swept out to sea. By hitch-

LARVA ON RIVER BOTTOM

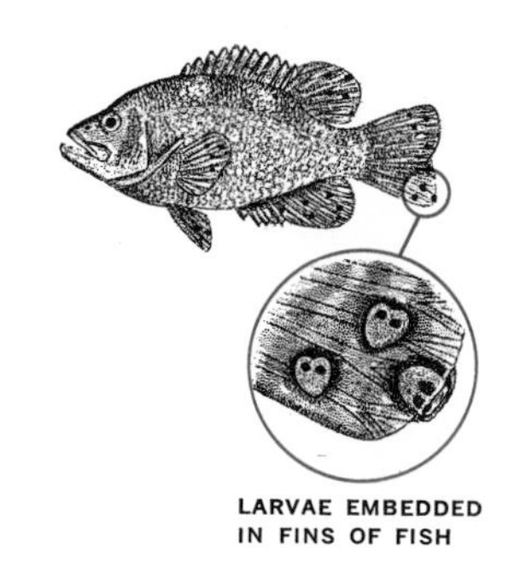

LARVAE EMBEDDED IN FINS OF FISH

MATURE LARVAE SETTLING ON RIVER BOTTOM

Of the millions of larvae that a single mussel may produce, only a few grow to maturity. When the tiny larvae are first released on the river bottom (*top*), the vast majority are eaten by predators or die of starvation. Some, however, attach themselves to fish (*center*) and spend several weeks as parasites. When the mature larvae finally drop from their hosts (*bottom*), still more perish because they land where bottom conditions are unsuitable.

ing rides on host fish, the larvae are distributed to suitable habitats throughout the river system.

Even so, mussels are far less abundant in most rivers than they were in the past. Many rivers now are too polluted for mussels to survive. In addition, damming of rivers has created lakelike areas where the water flows so slowly that it drops most of its load of silt. The result is a muddy bottom that is unsuitable for most kinds of mussels.

Life on the muddy bottom

Mussels are not the only creatures that dwell on the river bottom. Like smaller streams, rivers contain an assortment of stonefly nymphs, caddisfly larvae, mayfly nymphs, and other insects. But the species usually are different from those found in mountain streams, and for a very good reason.

Instead of jumbled stones swept clean of all debris by rushing water, great stretches of the river bottom are covered with mud, silt, and sand. Suckers, claws, and other devices for clinging to the bottom are useless in this constantly shifting habitat. Most insects that live here escape the current by burrowing in the bottom. Dragonfly nymphs lie half buried in the mud and wait in ambush for prey. Mayfly nymphs are equipped with flattened spadelike snouts and paddlelike forelegs for burrowing through the mud. *Tubifex* worms, small relatives of earthworms, also may be incredibly abundant on the muddy bottom.

The same situation occurs in slow-flowing streams that meander across level meadows and flatlands. Where rivers hurtle across rapids, on the other hand, you will find the usual assortment of black-fly larvae, water pennies, and other creatures characteristic of swift upland streams.

In areas of rivers where strong currents keep bottom deposits shifting constantly, life is so difficult that few animals survive. But in protected places where the current is slower and the bottom is more stable, or in shallows near the shore, insect populations may be enormous.

Yet without special equipment it is difficult to study life on the river bottom. Except near shore, the water in most

Where a slow-flowing river meanders across nearly level land, sand and silt are deposited on the bottom. The result is a set of living conditions far different from those of swift upland streams, where bottom rocks are swept clean of all debris.

The mating flight of damselflies is similar to that of their larger relatives the dragonflies. Grasped firmly behind the head by her mate, the female swings the tip of her abdomen around to the pocket under the male's abdomen and picks up some of the sperm capsules cached there. . . .

places is simply too deep for you to observe the animals firsthand. More often than not, the water also is muddy. Just as you had to rely on the activities of muskrats to detect the abundance of mussels, you must rely on indirect evidence to learn about other inhabitants of the river bottom.

The flight of the dragonflies

Many kinds of dragonflies pass their nymphal stages in rivers, just as others live in streams, ponds, or marshes. But even if you never see the nymphs, activities of the adults tell you that there must be dragonflies maturing in the river. As you drift downstream, glinting metallic-looking adults are certain to dart by from time to time or even to alight momentarily on the prow of your canoe. Occasionally you may spot two dragonflies linked in flight, like a miniature airborne train, yet maneuvering with perfect coordination. The two are mating.

The one in front is the male, gripping the female by the

top of her head with special claspers at the tip of his slender abdomen. Before joining, the male transferred sperm capsules from the tip of his abdomen to a pocket near its base. Now as they fly, the female doubles forward in a J curve to pick up the sperm and fertilize her eggs.

Now and then the pair lights on the stem of a cattail or some other plant near the river's edge, and the female attaches eggs to the stem just beneath the water. Or the male may release his grasp and wait nearby while the female lays her eggs. After she finishes, they join once more and fly off to seek another site for depositing still more eggs. When the nymphs hatch, they drop to the bottom and take up their predatory existence in the river muck.

In some species, the female simply scatters a trail of eggs across the surface of the water. Or she may swoop occasionally and touch the tip of her abdomen to the surface to deposit eggs. Damselflies mate in much the same way as dragonflies, except that the male clasps the female a bit farther back on her body and the female generally deposits her eggs in plant tissues by piercing the stems with a knife-like structure at the tip of her abdomen.

. . . The pair later alights on streamside plants, and the female, dipping her abdomen into the water, releases fertilized eggs. Taking to the air again, the damselflies repeat the performance until all the eggs have been fertilized and deposited in the water, a process that takes an hour or so.

Millions of mayflies

Mayfly nymphs also are abundant in most rivers, yet they go unnoticed until they emerge as adults. As you cruise quietly downriver, especially towards evening, you may notice a cloud of the adults flying up and down in unison above the water: the mayflies are enacting their mating ritual. Within a day or two, when their brief existence as adults has ended, the mayflies' spent bodies litter the surface of the water and accumulate on sidewalks and pavements in riverside towns.

Near the Mississippi and other great rivers, mayfly swarms sometimes are incredibly abundant. At Sterling, Illinois, for example, so many mayflies emerged from the Mississippi on July 23, 1940, that dead bodies piled up in heaps as high as four feet on one bridge. Throngs of dead mayflies made the pavement so slippery that trucks could cross only if they were equipped with tire chains to prevent skidding. According to local newspaper accounts, it took a crew of fifteen men in hip boots nearly two hours to clear the roadway of insects. Shovels alone proved inadequate for the task, and a snowplow finally had to be pressed into service to clear a path across the bridge.

Mayflies long have symbolized everything that is fleeting and impermanent. After spending months or years as aquatic nymphs, mayflies emerge in great swarms, mate, lay eggs, and die. Their life's work is completed within hours or at most a few days after their transformation to winged adults.

At first glance this half-inch-long midge looks like a mosquito, but it has no bite. Feathery antennae mark this one as a male. Like mayflies, midges gather in enormous mating swarms, in which the beating of millions of tiny wings produces a humming sound that may be audible at a considerable distance.

A swarm of midges

The midge family is another insect group whose abundance is revealed indirectly through the presence of large swarms of flying adults. Larvae of various species thrive in almost any type of water—in lakes, ponds, marshes, and streams. On muddy river bottoms they may be so numerous that their slender, threadlike bodies provide a major source of food for many fish. As many as fifty thousand per square meter—or more than thirty larvae to a square inch—have been observed in some areas.

Most midge larvae feed on bits of decaying material in mud and water, although some eat microscopic plants. The commonest kinds are called bloodworms, though they are not worms at all. The larvae live in tubes in the muck and continually undulate their bodies to maintain a flow of water that brings both food and oxygen.

When midge pupae have completed their transformation to adults, they emerge from the water and fly in swarms

A saucer-shaped mass of sticks cradled on the limbs of a dead tree is the nest of a great blue heron. The bird seems awkward as it comes in for a landing, yet herons are able fliers despite their size.

sometimes so dense that they resemble billows of smoke. The clouds of tiny insects that you occasionally notice hovering over trees or bushes beside the river are swarms of midges. Although they may be annoying, most species are harmless, since they are unable to bite.

Each individual in the swarm constantly moves up and down, gyrating in endless frantic circles above the water. Most of the fliers are males, which you can recognize by their large feathery antennae. Now and then a few females join the swarm, mate, and soon drop jellylike coils of eggs into the water.

But many never reproduce, for swallows swoop tirelessly over the surface of almost every river, skimming the water and then zooming high overhead as they scoop up thousands of midges in their wide-open mouths. Many other birds catch midges on the wing, as do dragonflies, robber flies, and other predatory insects. Bats also take their toll of flying insects, but they do not emerge from hiding until dusk; you are likely to see them only as shadows fluttering across the face of the moon.

River turtles

As you guide your canoe downstream—past a farm, beneath a bridge, and then into the forest once again—animals far more conspicuous than bottom-dwelling insects are likely to attract your attention. A great blue heron stalks with elegant grace through the shallows, searching patiently for the darting forms of fish. With a lightning-quick thrust of its beak, the bird snaps a young bass from the water and swallows it in a single gulp. Its smaller relative, the green heron, eyes you from a tangled willow thicket and then, with a hoarse *skeouw*, flaps lazily upstream. A water snake skims silently past your canoe, with only its head showing above the surface; then even that disappears as the snake pursues a luckless fish or frog.

Just ahead, several painted turtles are basking on a fallen log. At your approach they slip one by one into the water,

Treading the shallows with a slow and stately gait, the four-and-a-half-foot-tall great blue heron is the epitome of haughty dignity. But woe to the fish that comes within range of its daggerlike bill—a quick jab and the victim is snapped up whole. Although fish are the staple of the great blue's diet, this graceful wader is equally content to dine on frogs, snakes, crayfish, large insects, or even small mammals.

THE RIVER'S SOLID CITIZENS

Among the most ancient—and successful—of all reptile groups, turtles first appeared on earth about 200 million years ago. After 50 million years of refining and perfecting their unique boxlike body plan, turtles plodded through the next 150 million years of their existence with few additional changes in their basic anatomy. Secure in the protection of their bony shells, they saw the mighty dinosaurs come and go and witnessed the rise of birds and mammals. The earliest turtles were swamp dwellers, and many modern ones remain so. Others have taken to the woodlands, the ocean, and even the desert.

The snapping turtle is considered by many to be one of the ugliest and meanest-tempered creatures on earth. Its toothless beak is a sharp, powerful weapon that can be used with lightning speed.

*The painted turtle (*left*) is probably the most common species of American turtle. It spends a great deal of time basking in the sun, but it usually slides into the water before you can get near enough to observe it closely. The red-eared turtle (*above*) is frequently kept as a pet. Given good care, it may grow to a length of six inches or more, although its attractive markings fade with maturity. The softshell turtle (*right*) lacks the heavy armor of most of its relatives but compensates for its vulnerability with a razor-sharp beak and a quick temper. If you must, handle it with care; better still, don't handle it at all.*

making hardly a ripple. If the river is sluggish and muddy, it very likely harbors snapping turtles. Since they seldom bask in the sun as most other turtles do, they are rarely seen. These large, ill-tempered creatures eat almost anything, from ducklings and fish to dead plants and snails.

Along the Gulf Coast and the southern Mississippi Valley, you may also see the alligator snapping turtle, one of the largest fresh-water turtles in the world. These ungainly creatures grow up to a yard long and sometimes weigh 150 pounds or more. Lurking on the bottom, with their huge mouths agape, they snap up small fish that are lured within range by the constant wriggling of their pink wormlike tongues.

A fish's shape often reveals a great deal about its habits and habitat. Because its body is flattened from side to side, a sunfish would be practically helpless in a swift stream: each time it turned sideways, the current would sweep it downstream. But in the quiet, weed-grown shallows where it lives, its waferlike body form is a useful adaptation for slipping through dense stands of vegetation.

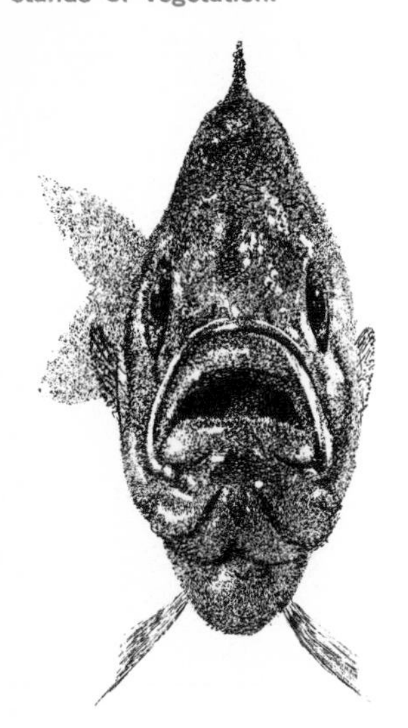

Through much of the central and southern United States, you may also see the softshell turtle, one of the few species of turtles that live primarily in rivers. Its nearly round and somewhat flattened body is covered by a flexible leathery shell. Although it sometimes basks on shore, more often it lies buried in sand in the shallows, with its long neck extended and just the tip of its curiously pointed snout projecting from the water. If a small fish, frog, or other animal comes too close, the turtle strikes out suddenly and snatches the victim with its sharp jaws. It is wise to study the softshell turtle from a distance, since it is apt to bite viciously when handled.

Fishes everywhere

Now and then the placid calm of the river's surface is broken by a sharp splash, then all falls silent once again. If you look back to the source of the sound, you see only a ring of ripples spreading across the water. But if you happen to be looking in the right place at just the right moment, you will see a fish rise from the water, then slap the surface with its tail as it goes under once again.

If the river is sluggish and muddy, the fish may be a carp. These giants of the minnow family often weigh twenty pounds or more. Their bodies are chocolate brown tinged with orange and are checkered with exceptionally large scales. Introduced to America from Europe about 1870, carp are a classic illustration of the hazards involved in importing exotic plants or animals to new habitats. They found living conditions in our rivers and ponds so satisfactory that they

have become a major nuisance in many waterways. Carp root in the bottom like pigs, eating plants, fish eggs, small animals, and almost anything else they find. In the process, they make muddy water even muddier, so that bass and other more desirable fish cannot survive.

If you nose your canoe in near shore, you may discover a shallow basin scooped from the sand or pebbles. A wafer-thin sunfish hovers over the basin, but at your approach it darts away. Half hidden in a tangle of weeds, the fish lurks nearby and returns to its station as soon as you depart. The depression in the gravelly bottom is the sunfish's nest; the male will continue to stand guard over it until all the eggs have hatched.

Catfish, or bullheads, probably live in the river also. These highly adaptable fish thrive in ponds, small streams, and large rivers as well, where certain species grow to great size. Even in muddy water they have no difficulty locating food. The whiskerlike processes that droop beside their mouths and dangle beneath their chins are supersensitive feelers that enable catfish to locate any food—plant or animal, living or

Colorful sunfish, small relatives of the bass, are common residents of rivers, streams, and ponds throughout central and eastern North America. Most of the smaller species are schooling fish, while larger ones tend to be solitary predators.

Catfish, or bullheads, of various species are well-known residents of nearly all American rivers. The channel catfish, the most widespread member of the family, may weigh as much as twenty pounds. But the blue, or Mississippi, catfish is the giant of the family, occasionally tipping the scale at a hundred and fifty pounds.

KILLIFISH

dead—that may lie on the bottom. Like sunfish, they excavate nests in the bottom and even protect the schools of tadpolelike young for a time after they have hatched.

But here there are no trout, no darters, no sculpins or other fish characteristic of swift upland streams. The river water is too warm for them, and other more heat-tolerant fish have taken their place. Largemouth bass, crappies, buffalo fish, gar, eels, and a number of others thrive in smaller rivers. In truly large rivers, even more spectacular fish begin to appear: gigantic paddlefish in the Mississippi River Basin, thousand-pound sturgeon lurking in the depths, enormous schools of salmon that migrate up the Columbia and other rivers to spawn. Each plays its role in the economy of the river, consuming plants or smaller animals and providing food in turn for still other river creatures.

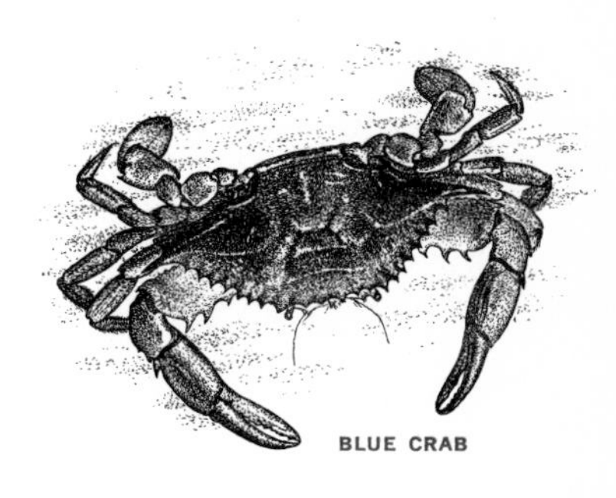
BLUE CRAB

The end of a journey

If time were no problem and you could continue drifting downstream indefinitely, your highway to the sea would undergo great changes. Other rivers would merge with yours, and your river would flow into still larger ones. Eventually you would smell salt in the air and gulls would become more and more numerous. Your river would be approaching the end of its journey, and just a few miles ahead it would lose its identity in the vastness of the ocean.

OYSTER

As a river approaches the ocean, its water becomes mixed with increasing amounts of salt water from the sea. Three of the many different kinds of animals adapted to life in this zone of mixing are shown here.

The mouth of the river, where fresh water mixes with salt water, is a fascinating zone that harbors its own unique community of plants and animals—a community so complex that a person could spend a lifetime unraveling its secrets. But for our purposes it is enough to note simply that the community of life changes where the river flows into the sea. As water becomes increasingly salty near the ocean, the familiar forms of fresh-water life gradually disappear and other more salt-tolerant forms take their place.

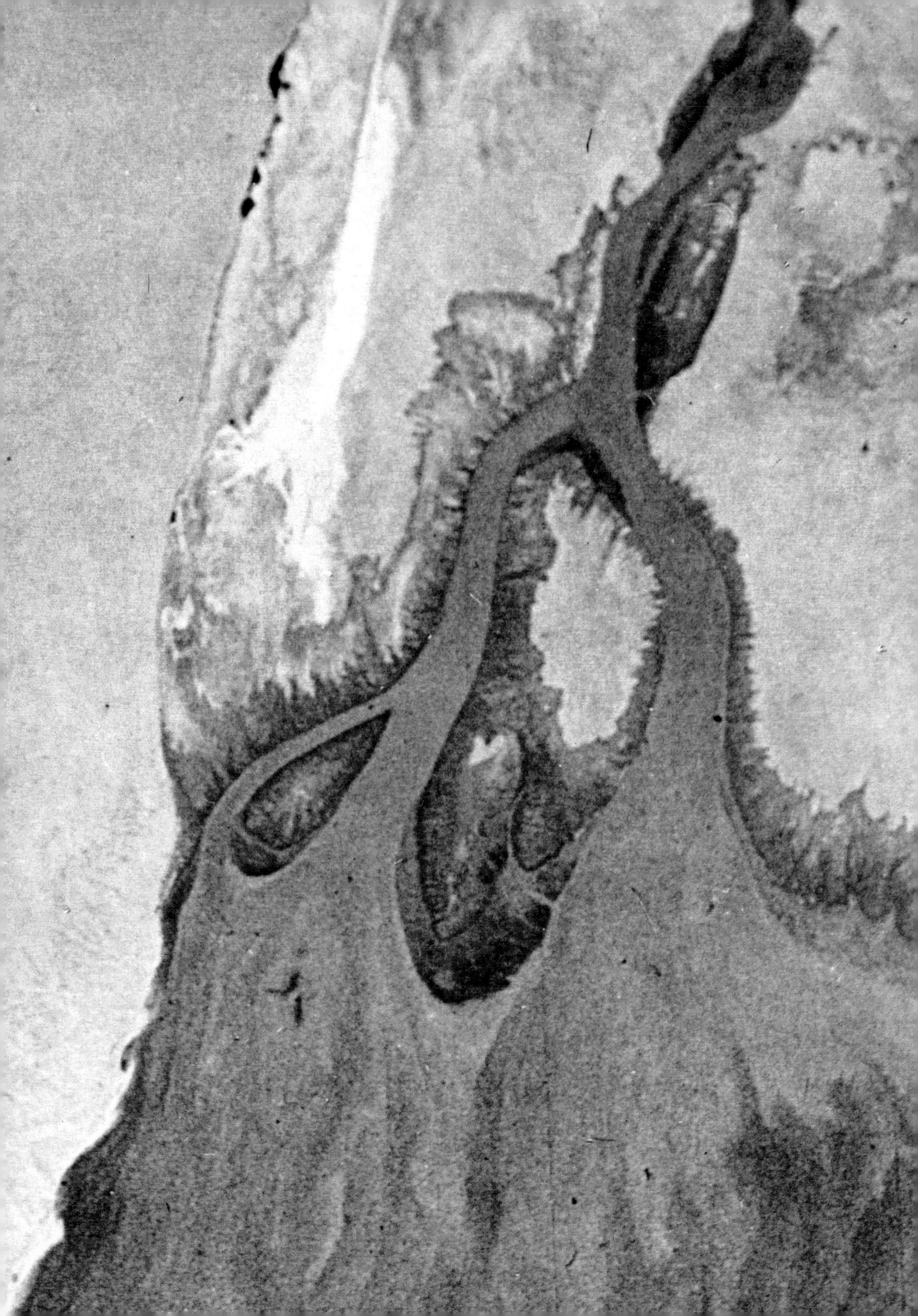

Instead of insects at the mouth of the river, the bay and tidal marshes swarm with crustaceans, such as crabs, shrimps, and many smaller forms. Instead of fresh-water mussels, there are oysters, razor clams, and salt-water mussels. Carp, bass, and sunfish gradually give way to killifish, flounders, croakers, and a host of other fish. Plovers forage along the banks, and pelicans, cormorants, terns, and other less familiar birds skim through the air.

The change from fresh-water to salt-water forms is gradual, of course. Some fresh-water animals can tolerate more salt than others and manage to survive closer to the sea. By the same token, some of the creatures of the sea can penetrate farther inland than others. But if you traveled far enough, you would finally reach an area where all the plants and animals were true salt-water forms. The seawater would no longer be diluted by the fresh water pouring from the river.

All the water that the river carried from countless springs, swamps, glaciers, and other sources hundreds of miles inland would at last have reached its final destination. Yet its journey would not be over, for water never ceases in its endless travels. No matter where the great ocean currents may carry the water from your river—to the Arctic or the Antarctic, to the Mediterranean Sea or the Indian Ocean—the water is bound to evaporate again, rise over the continents, and fall as rain once more.

After flowing some 1450 miles from sources high in the Rocky Mountains, the water of the Colorado River spills finally into the sea at the Gulf of California. Yet its journey is not over, for the ceaseless movement of all water from sea to land and back again to the sea is a cycle that has neither a beginning nor an end.

Rivers of Life

Streams and rivers are more than flowing waters that carry dissolved gases, minerals, and bits of sand and silt on a headlong course to the sea. In a sense they are factories. Unlike a steel mill or a furniture factory, however, a river does not convert raw materials into pipes or tubing or tables and chairs. The finished products of a river are communities of living plants and animals. The beds of cattails that fringe the shallows, the muskrats, the turtles, the fishes that school in the depths, all depend for their survival on the continuing health and productivity of the river.

The earth's supply of the basic raw materials of life is strictly limited. Only so much carbon, oxygen, nitrogen, and so forth, exist in the world. Yet the same basic reservoir of these materials has been used in the production of life for more than two billion years. This is possible only because the same substances are used again and again. The raw materials locked temporarily in the living leaves of a cattail, for example, are not lost forever from the great chain of life. A muskrat that eats the leaves converts them to animal flesh. When the muskrat dies, the processes of decay in turn convert the complex compounds in the animal tissues to

simple raw materials that are once again available for the growth of plants. Under natural conditions, the only limitation on this process of constantly recombining raw materials into new forms of life is the available supply of energy, for energy is expended at each step along the way. This energy comes from the sun.

The river factory

Like any factory, the river's productivity is limited by its supply of raw materials and its efficiency in converting those materials into finished products. If a stream is well supplied with raw materials, such as oxygen, carbon dioxide, and vital elements, and receives sufficient sunlight, it will be rich in plant and animal life. If any of the raw materials are scarce, if the river is polluted, or if it does not receive enough sunlight or organic materials, its output of living things will be low.

The basic forms of life in rivers, as in every habitat, are algae and other green plants. They are called *primary producers,* since green plants alone, of all living things, possess the ability to harness the energy of sunlight and transform simple, nonliving substances into living tissues. The remarkable process by which plants perform this feat is known as *photosynthesis.* Although scientists do not understand exactly how photosynthesis works, they do know that chlorophyll, the green coloring matter of plants, contributes to the process by absorbing energy from the sun. This energy powers a series of chemical reactions that combines carbon dioxide with the hydrogen in water molecules, to produce simple sugars and release oxygen. In simplified terms, we can describe the process in this way: Carbon dioxide plus water plus energy yields sugar plus oxygen. The sugar is used by the growing plant, and the oxygen remaining at the end of the process is given off into the air or surrounding water.

In the stream, as in any natural habitat, green plants perform the basic work of converting nonliving raw materials into living tissue. Some common types found in shallows and placid backwaters are yellow water lilies *(left)*, tiny floating duckweeds *(right)*, and stately six-foot-tall cattails *(far right)*.

Green plants everywhere

In sluggish backwaters and in shallows near the shore, plants of all shapes and sizes spread their leaves to the sun. Cattails, bur reed, arrowhead, pickerelweed, and many others form dense stands of greenery in the shallows. Water lilies dot the surface with glossy floating leaves that are anchored firmly in the bottom muck.

In the South, many rivers are blanketed with water hyacinths, curious plants that normally float freely on the surface. Pontoonlike air bladders at the bases of the leaves keep the plants afloat. Despite their beauty, these imports from South America are considered a nuisance; water hyacinths often cover the surface so completely that boats can scarcely get through. As a result, government agencies spend millions of dollars trying to eradicate or control these amazingly prolific weeds.

The finely divided underwater leaves of water crowfoot contrast sharply with the broad floating leaves of a pondweed. Because its feathery leaves offer little resistance to currents, this white-flowered relative of buttercups can flourish in rapid streams where most rooted plants would be torn to shreds.

A beautiful nuisance accidentally imported from Brazil, the water hyacinth is also known as "water orchid." Florida taxpayers call it the "million-dollar weed" because of the great expense involved in clearing their waterways of this fast-spreading floater.

Springs and cold brooks also may be clogged by lavish growths of plants such as water cress or duckweed. Or, if the current is not too strong, the open water of a stream may be filled with graceful stalks of water crowfoot, a kind of buttercup. Although the crowfoot's surface leaves are broad and delicately scalloped, its underwater leaves are divided into slender feathery lobes; broad underwater leaves would soon be torn to shreds by flowing water. *Fontinalis*, the fountain moss, also sends out slender stems from its foothold on the rocks in swift water.

Although rooted plants are the most conspicuous vegetation in any river or stream, they play a relatively minor role in the river's productivity. The most numerous plants of all are algae, simple plants that have neither roots, stems, nor leaves.

The green film of life

Some of the algae are fairly obvious to any observer. Stagnant water is often filled with tangled, slimy strands of threadlike *Spirogyra.* In moderately rapid streams, masses of the alga *Cladophora* may be anchored to the rocks, with hairlike filaments of the plants billowing downstream for several feet. Or the rocks may be studded with small bluish-green pads of the alga called *Nostoc.*

Most algae are too small to be seen without a microscope. Thousands of different kinds grow in an unimaginable variety of forms, but the most numerous algae are the tiny one-celled types called diatoms. Each diatom is encased in a delicate shell composed primarily of silica and consisting of two valves that fit together as a lid fits a box. The glasslike shell of every species has its own particular shape and uniquely sculptured surface pattern.

Despite their small size, microscopic plants are so abundant that they may visibly color the water. Millions upon millions of them make up the greenish-gray scum that sometimes drifts lazily on the surface of still waters. More important, algae, along with bacteria and microscopic animals, are the substance of the slick brownish film that coats nearly every object in the stream—rocks, logs, plant stems, even the shells of living turtles. This thin film of life is the very basis of the stream community. Of all plants in the stream, algae perform the greatest work of capturing the energy of the sun in living cells and making it available to other inhabitants of flowing water.

Beneath a microscope, the streaming mass of algae on the opposite page might be identified as *Spirogyra* or *Cladophora,* two kinds that are common throughout the United States. To the unaided eye, both appear as formless strands of green, yet their internal structures are distinctive.

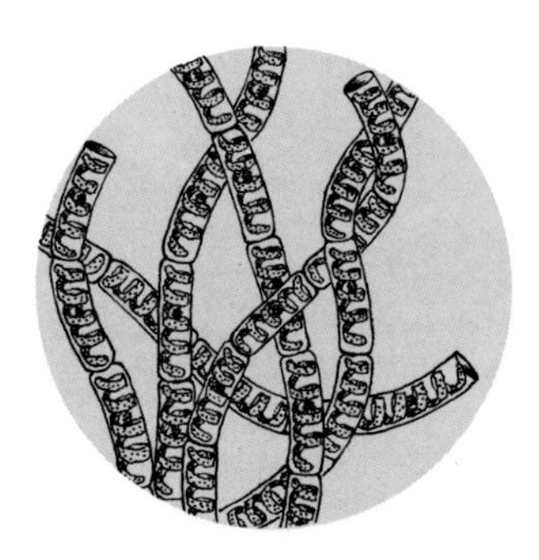

Spirogyra

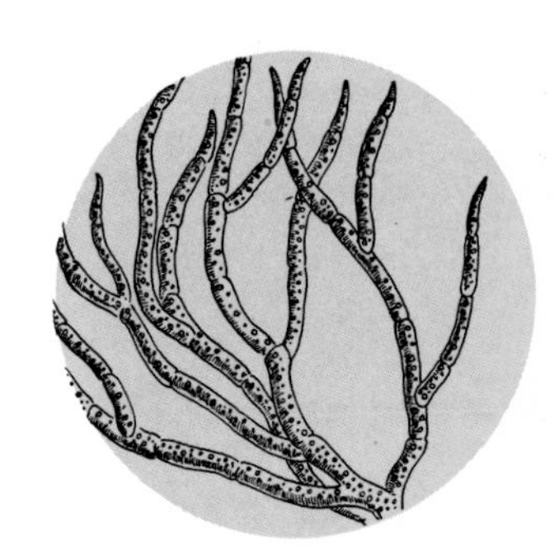

Cladophora

Along the assembly line

Animals cannot manufacture their own food. Unlike plants, they must obtain energy and nutrients secondhand, either by eating plants or by eating smaller animals that have fed on plants. This flow of energy and material from one form of life to another is known as a *food chain.* It usually involves four or five links, or transfers of energy and material.

Whether growing as a slick film on rocks and other surfaces or as long hairlike filaments, like the kind shown here, algae of many species are the river's most important primary producers. Under a microscope each of the tiny plants is revealed to be a beautiful and intricate factory for producing food materials.

First come the producers, the algae and other green plants. Then there are first-order consumers, animals such as mayfly nymphs or snails, which rasp algae off surfaces in the streams. These small *herbivores*, or plant eaters, in turn are eaten by second-order consumers, the *carnivores*, or flesh eaters, such as sunfish. A third-order consumer, such as a bass, may then eat the sunfish, while the otter or the fisherman who eats the bass for dinner becomes a fourth-order consumer.

Or the bass may die a natural death. In this case the energy and nutrients stored in its body will be released by *decomposers*, such as bacteria, or consumed by *scavengers*, such as crayfish.

Examining a single food chain reveals only part of the story, however, for hundreds of food chains exist in any living community. Most of the chains are interconnected. Instead of being eaten by a man, the bass might be snatched up by a heron or captured underwater by a mink. The mayfly nymph might escape the jaws of hungry fishes, only to emerge as an adult and be scooped from midair by a bat or a swallow. Because all the simple food chains in a community are interconnected in this way, the organization of the community is more accurately described as a *food web*.

Yet the basis of this complex system of food relationships is always the same. In streams, as in every other natural habitat, the primary building blocks are green plants; they alone are able to unite the energy and materials that are eventually reassembled to form every other living thing in the community.

Every plant and animal forms a link in one or more food chains, series of feeding relationships by which the food produced by green plants is distributed throughout a community. In this simple five-link food chain, for example, microscopic algae are the primary producers of food. Herbivorous mayfly nymphs graze on the algae and are eaten in turn by carnivorous sunfish. Bass feed on the sunfish, and otters eat the bass. Thus, even the largest of the stream's predators depend ultimately on the primary productivity of green plants.

Food that floats

In many streams a good deal of food material is washed along with the current. This *drift,* as it is called, may be so abundant that it provides a major food source for certain animals. Black-fly larvae spread their feathery mouthparts to the current in order to strain out passing bits of food. Larvae of the caddisfly *Hydropsyche* build fine-meshed silken nets on the faces of stream-bed rocks in order to filter out the drift.

Part of the drift consists of plants that have been torn free of their roots, plus bits of the decaying bodies of dead plants and animals. Following a flood, the water may be filled with small animals that have been ripped free of their holds on the bottom by exceptionally strong currents.

The major component of drift is likely to be *plankton,* minute plants and animals that normally float free in the water. Most of the plant plankton is composed of diatoms and minute, one-celled blue and blue-green algae. In flowing water, however, a great many of the floating algae usually are individuals that have been dislodged from the bottom by the current. One-celled protozoans and rotifers are the most important animals in plankton. As with microscopic plants, however, the majority of these tiny animals inhabit the film of scum that coats rocks and other surfaces.

Since currents sweep the floaters relentlessly toward the sea, plankton in streams seldom builds up to the great numbers found in the still water of ponds and lakes. In a river, plankton is always most abundant just downstream from places where the flow has been interrupted by lakes.

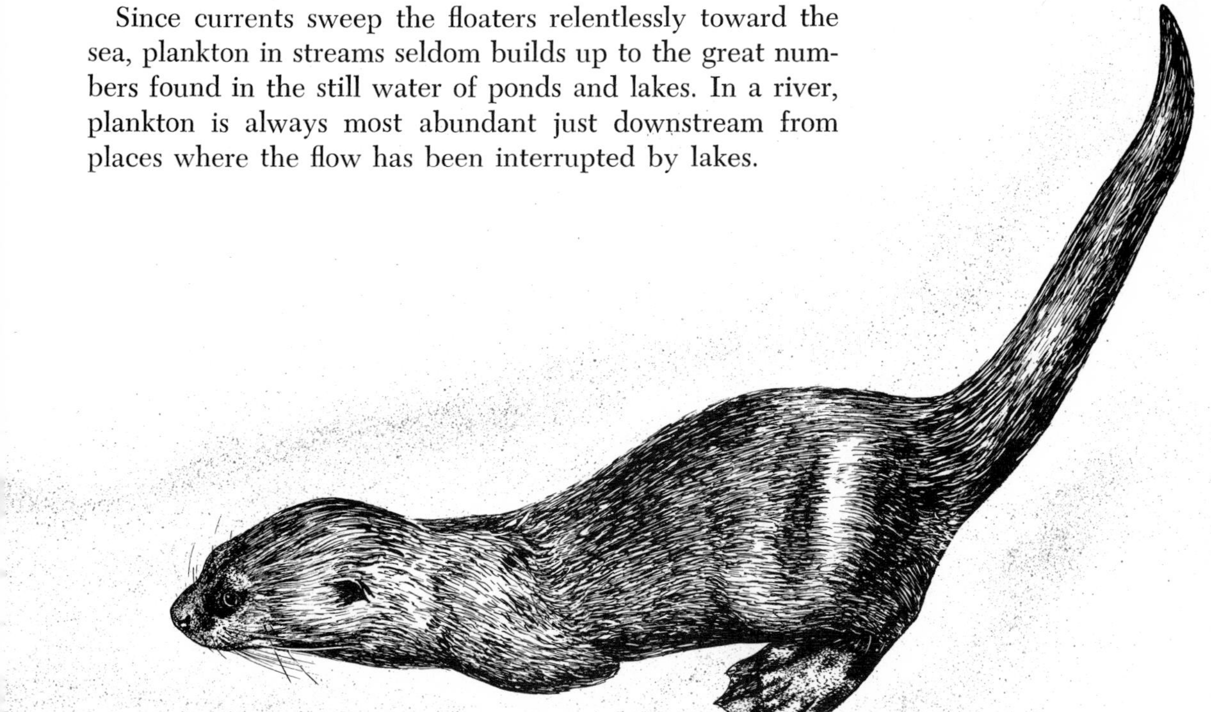

Leaves, insects, and other materials that fall into the water make an important contribution to aquatic food webs. Some are eaten directly by stream animals. Others are broken down by decomposers, releasing nutrients that support the growth of green plants.

Food that falls in

The land also makes a contribution to the stream's food web. Moist soil at the water's edge frequently is covered by dense growths of willows, alders, and hundreds of other trees and shrubs. The trees may shade the stream completely and thus help keep the water cool. Even more important, the leaves that fall from overhanging branches are a major source of food and organic material in many streams and rivers. Besides being eaten directly by small animals such as mayfly nymphs, the dead leaves that accumulate among rocks and fallen branches are decomposed by bacteria and fungi, releasing vital nutrients into the water. As an added benefit, the tangles of debris provide excellent hiding places for hellgrammites and other animals.

The overhanging branches of willows and other trees also produce a shower of aphids, leafhoppers, and other insects that fall continually to the surface and drift downstream. Shake the limb of a tree and you may see a trout rise to the surface, snatch an insect that has fallen in, then swirl from sight beneath an overhanging bank. Insects such as mayflies also form an important part of the fish's diet. After mating, the insects die and fall to the surface of the water, providing a feast for the fish.

Food that falls into the water is so important, in fact, that trout fishermen are able to deceive the fish with dry flies. Some of these lures, such as the "royal coachman," do not resemble any living insect. But many, made of feathers, bits of hair, and other materials, are faithful copies of mayflies, caddisflies, and other insects.

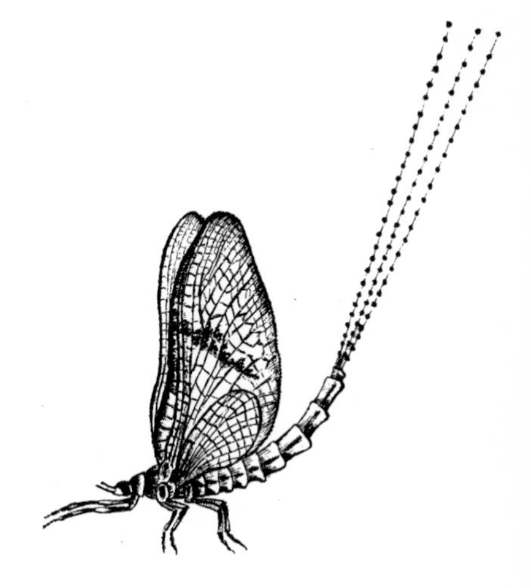

Mayflies and other insects that fall to the surface of rivers and streams are a favored food of trout and other fishes. Aware of this fact, generations of fishermen have used feathers, fur, wax, and lacquer to create colorful barbed facsimiles of living insects.

Measuring productivity

How can you measure the productivity of a stream or river? Scientists have sought answers to this question for many years. Knowledge of the amount of life a stream is able to support, for example, may be important in establishing fishing regulations on a stream.

One method is to conduct a creel census—a count of the number and types of fish caught in a single day, season, or year. Unfortunately, the results depend on such things as the skill of the fishermen and the kinds of bait and equipment they use. Also, a creel census tells you nothing about the stream's output in terms of other animals, such as ducks, frogs, or nongame fish.

A better approach to measuring productivity is to sample the organisms that fish eat. A standard device used for this purpose is the Surber sampler, named for the man who developed it. It is used for collecting all the animals found within a square foot of riffle bottom. Riffle-bottom areas generally are much more productive than the stream's pool-type bottoms.

The Surber sampler consists of a one-square-foot frame connected to a second frame which holds a net. The operator sets the open frame on the riffle bottom; the net connected to the upright frame is swept downstream between his legs. One by one he lifts each stone and pebble in the square foot of stream bottom enclosed by the open frame and dislodges all the animals living there. As the current carries the animals downstream, they are caught in the net.

The Surber sampler is a device used for measuring a stream's productivity of food organisms. An open square-foot frame marks off the area of stream bottom to be sampled. A net on the downstream side collects creatures that the operator dislodges from stones in the sample area.

By counting and weighing the catch from a series of samples, a biologist can estimate the stream's productivity in terms of animal life per square foot of riffle bottom. In a rich stream, the number of animals usually exceeds two hundred, or one-tenth of an ounce, per square foot. Sometimes the number of animals may be as high as four hundred. A poor stream, on the other hand, yields fewer than a hundred animals, or less than five one-hundredths of an ounce per square foot. Once he has a measure of the stream's richness in fish-food organisms, the biologist can make at least a rough estimate of the number of fish the stream can support. Of course, other factors, such as the availability of shelter and spawning areas, also must be taken into consideration.

A major drawback to using a Surber sampler is the fact that it measures only the "standing crop," the number and weight of animals present at the time the sample is taken. Actually, many aquatic insects go through two or more reproductive cycles in a single year, while others require two or three years to complete a single cycle from egg to adult. Since insects form the bulk of the animal life on a riffle bottom, a better measure of productivity would be the "annual crop," the total weight and number of insects produced over the course of a year.

To arrive at a figure for the annual crop, we would have to know the complete life cycle of every animal found in the sample. Unfortunately, this information is known for relatively few of the thousands of kinds of plants and animals that live in rivers and streams. With a little equipment and a great deal of patience, nearly anyone can make a valuable contribution to science by rearing stream insects through their complete life cycles and by finding out which adults emerge from which larvae.

Primary productivity in Silver Springs

An entirely different method for measuring productivity was used in a study by Howard T. Odum, an ecologist at the University of North Carolina. Instead of measuring a stream's richness in fish, or in animals eaten by fish, Odum worked from the very base of the food chain. In a study that took several years to complete, he calculated a stream's primary productivity of green plants.

Where the casual observer sees only water swirling over rocks in a fast-flowing stream, closer examination reveals a flourishing community of plants and animals. Measuring a stream's productivity of fish-food organisms like these helps biologists estimate the stream's ability to support fish.

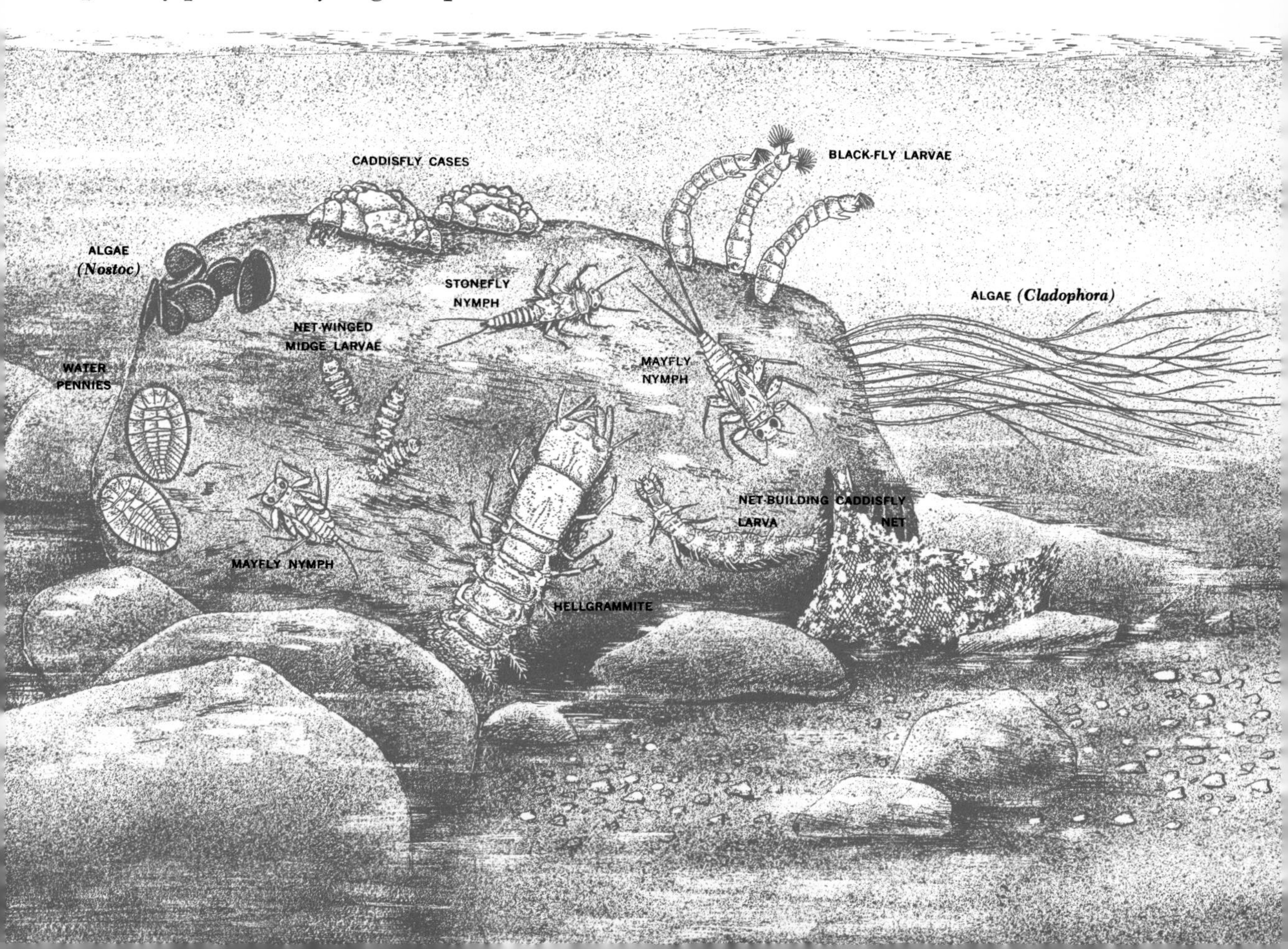

Odum conducted his study at Silver Springs, in north-central Florida. In this famous area a sparkling stream winds between banks lined by lush forests of oak, sweetgum, and other trees. Deer slip now and then to the water's edge to drink, and the forest is alive with birdsong. But the focal point of the area is the source of the stream, a huge artesian spring of the type common in this limestone region.

No mere puddle oozing from the base of a cliff, the spring forms a large, deep pool of crystal-clear water that bubbles up through crevices in its depths. Every day, year in and year out, 300 million gallons of water at a constant temperature of about seventy-four degrees pour from the spring. The stream that flows full-fledged from this source—really a small river—is so richly productive that thousands of tourists come each year to marvel at its teeming life.

Cruising along the river in a glass-bottom boat, tourists look down at dense stands of fresh-water eelgrass. The slender ribbonlike leaves ripple like streamers in the gentle current. Turtles hover in the eelgrass, nibbling on the tender

A two-foot-long bowfin glides through the eelgrass at Silver Springs, Florida, as it searches for insects, small fish, and other prey. This living fossil is the sole surviving member of a primitive group of fishes that flourished during the age of dinosaurs.

leaves, while great schools of fish—mullet, sunfish, large-mouth bass, bowfins, and others—flit like streaks of silver through the water. Huge gar, long-snouted fish covered with hard bony scales, lurk in the shadows. Now and then a gar darts from its hiding place and, with a lunge, snaps up one of the smaller fish. The water swirls momentarily where the great fish turned near the surface, and then the stillness returns.

Impressed by the obvious richness of this habitat, Odum and his co-workers decided to calculate its productivity. They found that the major primary producers were not the beds of eelgrass, but the dense coating of algae that encrusted every rock and blade of eelgrass in the stream. Mats of diatoms ensnared in tangled filaments of green and blue-green algae formed a slick coating of living plants on nearly every surface.

Standing sometimes ankle-deep in the mucky ooze on the bottom, the men used knives to scrape the algae from every inch of surface in carefully selected sample areas. When

they dried the material and weighed it, they found more than a pound of algae growing on each square yard of surface sampled. Just as with figures obtained by means of a Surber sampler, however, Odum's initial studies measured only the standing crop.

For really accurate figures, he needed to know the annual cycle of production. As might be expected, Odum found that productivity varied from month to month. Changes in productivity depended on such things as seasonal changes in sunlight and day-to-day variations in cloud cover and temperature. On the basis of a series of measurements, he calculated that an average of half an ounce, dry weight, of algae was produced per square yard of rock and plant surface per day. Over the course of a year, primary productivity was 12.85 pounds per square yard. Odum thus showed that productivity per unit of area in this extraordinarily rich habitat is greater than the productivity of many cultivated crops.

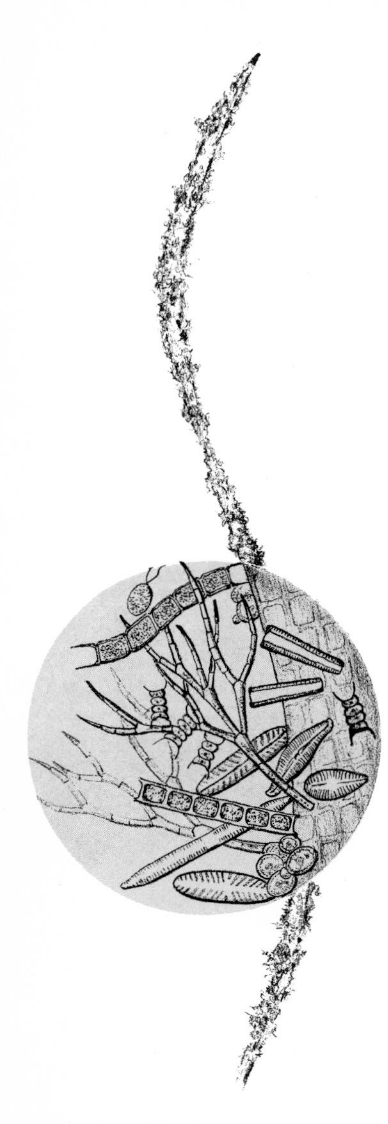

Under magnification, the coating of fuzz on blades of eelgrass at Silver Springs, Florida, is seen to be a tangled assortment of many different kinds of algae. Ecologists at Silver Springs found that these encrusting algae are far more important as primary producers of food than is the eelgrass itself.

The production pyramid

The surprisingly lush growth of green plants at Silver Springs obviously was sufficient to support a rich community of animals. Through a series of further studies, Odum calculated productivity figures for each link in the food chain. For these figures he concentrated strictly on the standing crop of plants and animals present at any one time. The total standing crop of primary producers, this time including both the fresh-water eelgrass and its encrusting layers of algae, averaged twenty-six ounces per square yard. At the next level were first-order consumers—snails, turtles, mullet and other plant-eating fish, and a variety of aquatic insects. These herbivores averaged one and two-tenths ounces per square yard. Second-order consumers, such as sunfish, catfish, predatory beetles, and other small carnivores, came to thirty-five one-hundredths of an ounce. Third-order consumers, primarily largemouth black bass and gar, averaged only five one-hundredths of an ounce per square yard.

You may wonder about this startling decline in the weight of organisms present at each level on the food chain. Why should spectacular twenty-pound gar seem so unimportant compared with the film of microscopic algae? In the first place, there are far fewer gar than algae in the stream. Just

Productivity studies at Florida's Silver Springs yielded dramatic evidence of the rapid decline in weight of organisms per unit area at each successive level on a food pyramid. The smaller pyramid *(below)* is drawn more nearly to accurate proportions in order to show the enormous base of plant producers necessary to support the community of animals at Silver Springs.

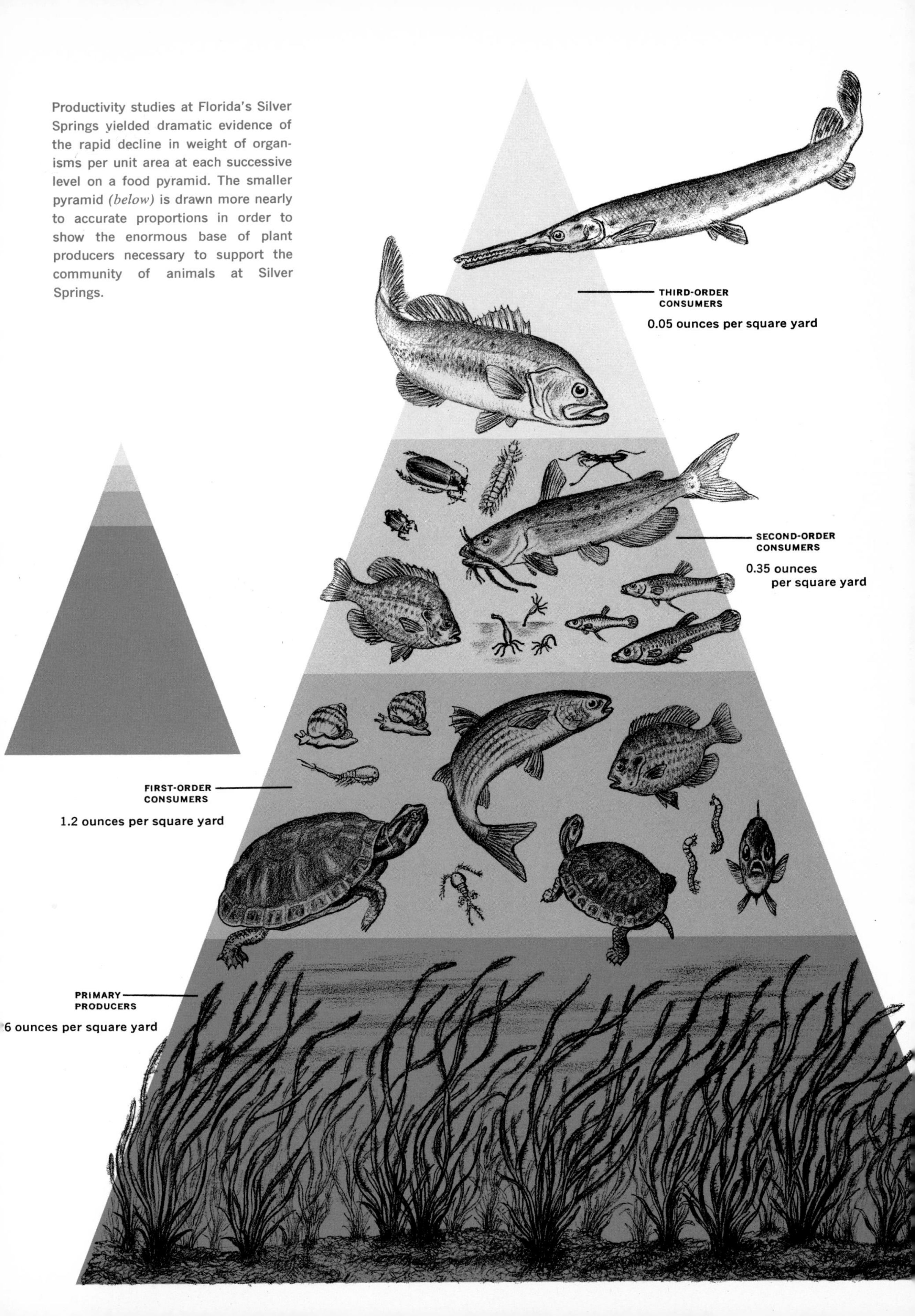

as it takes millions of one-celled algae to keep thousands of grazing insects alive, it takes thousands of insects to support just a few insect-eating sunfish. Gar, in turn, will be far less common than their prey, sunfish. This rapidly declining number of animals on each level along a food chain, described as a pyramid of numbers, can be observed in almost any habitat.

The same phenomenon can be expressed in terms of a production pyramid. Just as a great deal of energy from the sun is never used by plants, a great deal of energy escapes at each step along a food chain. Animals, in the first place, are relatively inefficient at converting their food into the substance of their own bodies; a great deal of energy is lost in the form of heat. Energy also is consumed by every activity or living process—walking, swimming, breathing, reproducing, and so forth. The weight of organic matter, and therefore the amount of energy stored in living bodies, thus is smaller at each step along a food chain. In addition, some of the energy at each level in the pyramid is locked up in organisms such as blue-green algae, which are eaten by very few animals. Instead of supporting higher-level consumers, they simply die and are broken down by decomposers.

In practical terms, this constant loss of energy means that a given base of plant producers in a stream, for example, can support just so many fish and no more. Or so it would seem. In terms of productivity of game fish, such as trout, the situation usually is much more complicated.

To catch a trout

Picture a trout fisherman standing hip-deep in the swirling waters of a mountain stream. The only sounds are the gurgling of water, the chant of vireos in the treetops, and an occasional swish as the fisherman casts a dry fly on the surface of the stream. Before long he hooks a trout, and after a brief struggle lands his prize with a skillful swoop of his net.

Picture the fisherman. But do not expect to find him alone. More often than not the stream will be lined by dozens,

Hip-deep in the sparkling water of a placid pool, a fisherman finds solitude as he tries his luck at the famous "Punch Bowl" on Oregon's Eagle Creek, a tributary of the Columbia River.

or perhaps even hundreds, of eager fishermen. And no wonder. At last count, nearly twenty and a half million fishing licenses were issued in the fifty states in a single year. It takes a lot of fish and a good many miles of stream to keep all of them happy.

A frequent solution to this problem is simply to ignore the natural productivity of trout streams. Instead of complete dependence on the stream's natural ability to produce fish, the water is stocked with trout raised in fish hatcheries. In many cases, the trout are full-grown when released. Fishermen stand elbow to elbow along the bank, eager to haul the fish out as rapidly as they are dumped in.

It would be far more efficient simply to hand each fisherman his quota of trout, yet the practice continues in many areas. Despite the hundreds of thousands of trout that are dumped into streams each year, many fishermen in heavily populated areas are still dissatisfied. Although fishing seasons and catch limits are carefully regulated, most of the fish are caught by the most skillful fishermen; hundreds of other anglers still go home with empty creels.

Trout are the most popular of all fresh-water game fish, but the natural supply often falls short of the fisherman's demand. A partial solution to this problem is to stock streams with hatchery-raised fish. A better method is to find ways of increasing the natural productivity of streams.

More fish for fishermen

Convinced that there had to be a better answer to the question of how to produce more trout for fishermen, the late Paul R. Needham, a zoologist at the University of California at Berkeley, and his students spent ten years studying the problem. After Needham's death, the results of his careful

study were compiled and analyzed by one of his students, Donald Seegrist.

In the first place, Needham knew that a great many hatchery-bred trout die as a result of the abrupt change in living conditions when they are released in streams. He realized, furthermore, that they compete with wild fish for the limited supply of food in the stream. But would it be possible to prove that a stream under natural conditions is capable of producing as many, if not more, catchable trout as a stream that is artificially stocked?

Needham and his co-workers took to the hills to find an answer. Their laboratory was a seven-mile stretch of Sagehen Creek, a rushing mountain stream high in California's Sierra Nevada. Several times in the 1930s and then from time to time until 1951, the stream had been stocked with rainbow, brown, and brook trout. During the ten years of the study, no more fish were planted in the stream. The study involved only natural propagation by wild fish.

Needham's first task was to find out exactly how many fish actually lived in the seven-mile stretch of stream. This was more easily said than done. Although there are many ways of sampling fish populations, none is as efficient or as accurate as the drastic method Needham used. Each August, he diverted sections of the stream into other channels, then pumped every drop of water from the by-passed sample sections of the original stream bed. In this way he was able to locate every single fish in the sample sections. The fish were counted, weighed, and measured, then returned to the stream when it was restored to its original channel.

University of California biologists spent a decade studying the ecology of Sagehen Creek, a Sierra Nevada trout stream. Once a year they pumped sections of the creek dry in order to obtain an accurate estimate of the number of fish living in the stream.

On the basis of his samples, Needham calculated the total fish population for the entire seven-mile section of stream. Over the ten years of the study, the stream averaged more than nine thousand trout under four inches in length, four thousand fish four to six inches long, and about two thousand over six inches. The total—more than fifteen thousand fish—obviously represented a great deal of potential sport for fishermen. Even more surprising was the fact that, despite heavy sport fishing, the numbers did not drop significantly over the ten-year period.

This remarkable stability was no reflection on the skill of California fishermen. Quite the contrary, in fact, for Needham discovered through careful creel censuses that fishing in Sagehen Creek was very good indeed. Over the course of each season, anglers averaged about three trout each per day. Fishing success, moreover, remained good throughout the season. The early birds did not get all the fish.

Although each fisherman was limited to a total catch of fifteen fish, his success in catching the limit depended mainly on how much time he spent. Experts caught the limit in as little as two and a half hours, while average fishermen needed about nine hours. Despite heavy fishing by a good many contented fishermen, the stream was not fished out. Many fish eluded the anglers and remained to grow and repopulate the stream with a new generation the following year. Needham found, for example, that one twenty-six-inch, seven-pound brown trout—a real prize—escaped the fishermen from 1954 through 1959.

So many fish—and no more

In all, Needham discovered that the trout population in Sagehen Creek averaged 150 pounds per mile. In the course of each season, fishermen were able to catch 38 pounds of fish per mile of stream—about a quarter of the total—without causing the population to drop the following year. Under existing regulations, the stream was a self-sustaining fishery. But what if more trout had been stocked? Would fishing have been even better?

Probably not. In order to thrive, fish need food, shelter, space to live in, and good spawning conditions at the right time of year. Other things being equal, fish will grow in direct proportion to the amount of food available. In experimental ponds, it has been found that if a given amount of food will produce two hundred pounds of fish per acre and a thousand fish are stocked, each fish grows to a weight of one-fifth of a pound. If two thousand fish are stocked, each will weigh only one-tenth of a pound. The pond still yields two hundred pounds of fish, but the fish are runts. This natural limit on the amount of life a stream or any other habitat can support is described as its *carrying capacity.* The proportion of its population that can safely be removed year after year without destroying its ability to maintain a stable population is called the *sustained yield.*

Needham demonstrated that Sagehen Creek has a carrying capacity of 150 pounds of trout per mile of stream and a sustained yield of 38 pounds. Artificial stocking would simply have resulted in overcrowding and smaller fish.

In streams where fishing pressure is exceptionally heavy, stocking may be the only means of maintaining fish populations. But Needham's study suggests that, in the long run, the best way to provide better fishing is by increasing the stream's carrying capacity: by increasing its potential productivity of green plants and other food, by providing more shelter or better spawning areas, and so forth.

Plump rosy trout, however, are only one value of rivers and streams. Besides yielding plants and animals, rivers also enter into the life of the surrounding area. In a sense, irrigated crops and even people are part of a river's productivity.

A great deal can be done to alter physical conditions in a trout stream in order to improve the stream as a trout habitat. Providing shade to keep the water cool, preventing bank erosion to keep the water clean, and providing better feeding, shelter, and spawning areas — all such measures yield dividends both in quantity and quality of fish populations. With the help and advice of a professional biologist, nearly anyone can build the simple, inexpensive devices illustrated here.

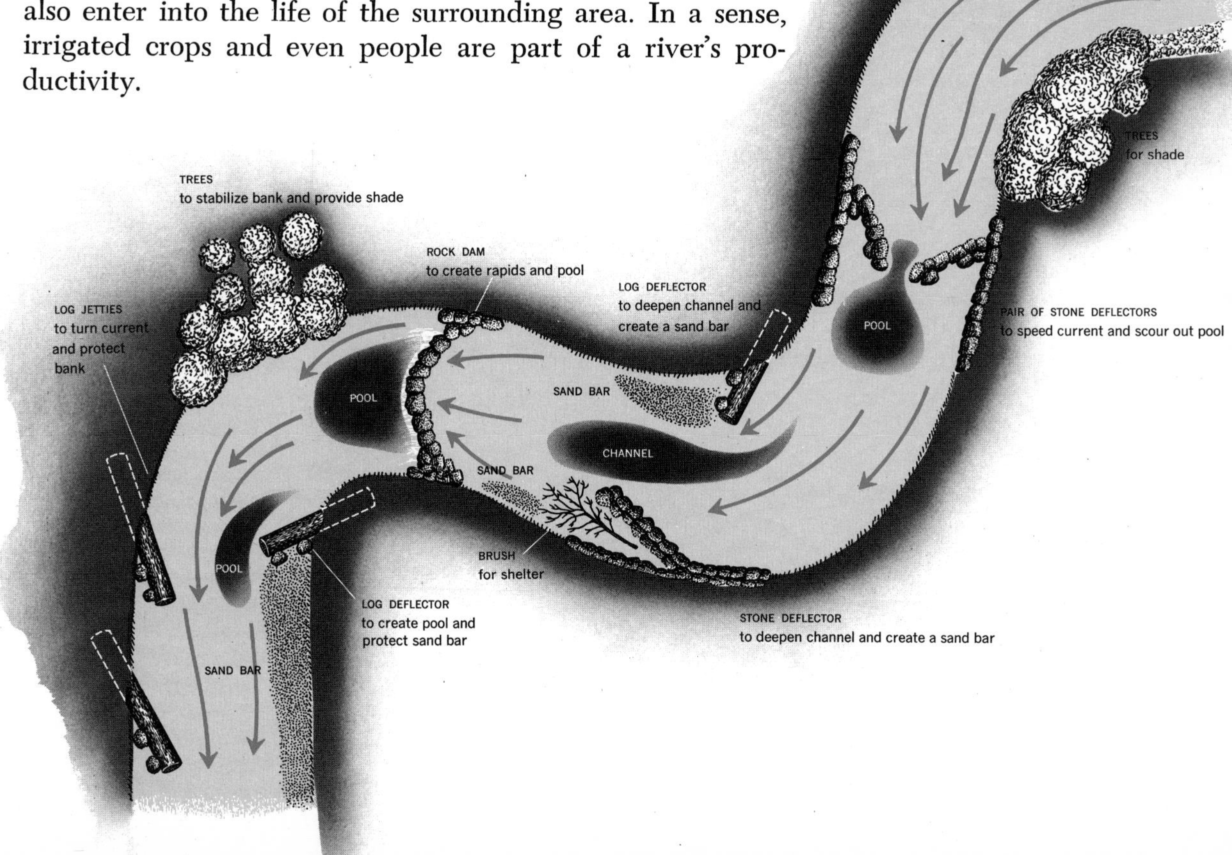

Rivers and people

Water is as essential as air to plants and animals. It makes up over seventy percent of the body weight of a man, and even more in some other animals—ninety-six percent for the jellyfish, for example. Since all the chemical reactions of life processes can take place only in the presence of water, no living thing can exist without it. A moderately active man in the United States must drink about five and a half pints of liquid a day. But beyond this basic biological requirement, we need water for many other purposes.

As human population swells, as cities grow larger, and as we learn to use water for more and more things, our needs for water increase each year. In ancient villages, each person used 3 to 5 gallons of water a day; in a modern city, each inhabitant uses 130 gallons a day. For all our cities, this comes to a total of about 16 billion gallons a day—and even this incredible figure does not include the staggering amounts used by industry. For instance, 110,000 gallons of water may be used in producing a ton of steel, and a million gallons is required to produce a thousand barrels of aviation gasoline. A large paper mill uses more water each day than does a city of fifty thousand people. Although this water is not destroyed, it may be evaporated or contaminated so that it is not immediately available for other uses.

To describe the situation in another way, our population has more than doubled since 1900, but per capita use of water has increased by more than four times. In some sections of the country, the increase has been even greater. Over a fifty-year period the population of Texas tripled, but the total use of water increased by seventy-one times. The largest single use of water in the United States today is for irrigation. This accounts for 75 to 100 billion gallons a day, or about half the fresh water we use annually. Naturally, the greatest use of water for irrigation is in the very areas where water is most scarce. The thirty-one eastern states, with a population of 128 million, use an average of 80 billion gallons of water a day. The seventeen arid western states have

Water sluicing down an irrigation ditch in the arid Southwest is put to work for man, transforming semidesert country into fertile farmland that yields bumper crops of sorghum and other agricultural products. Irrigation, the largest single use of water in the United States, consumes nearly a hundred billion gallons a day.

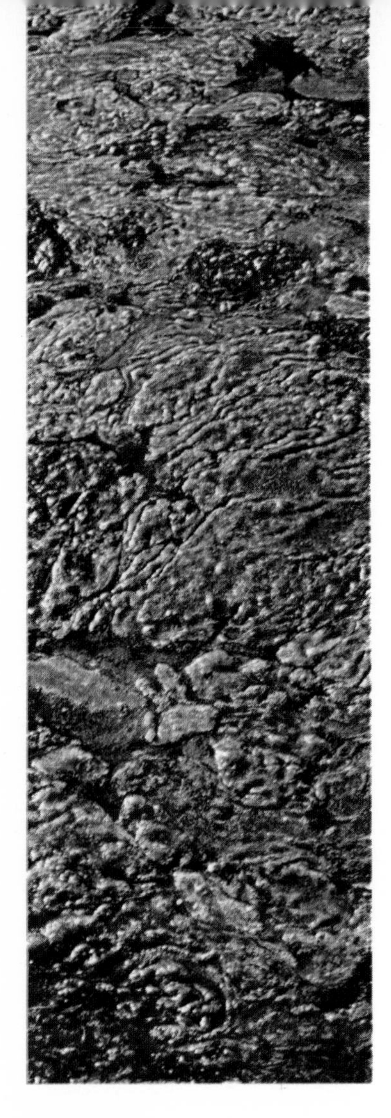

An explosive growth of algae is usually a sign of a sick river, one that has been polluted by an overdose of organic wastes. Few animals can survive beneath this smothering blanket of green.

only about one-third as many people (37 million). Yet, because of irrigation needs, they use even more water than the thirty-one eastern states—on the average, about 85 billion gallons a day.

As population continues to grow at an explosive rate, the situation can only get worse. A Senate committee has predicted that by 1980 the American people will be using 650 billion gallons of water a day. This is the absolute limit of all the fresh water now available in our lakes, streams, and reservoirs.

Most experts agree that the best solution to this dilemma is to increase our ability to reuse water. For water, as we have seen, is never really consumed. As it flows past cities or factories, water is removed from the river, used, then poured back into the river to be reused downstream. Great rivers thus are literally the lifeblood of the continent and all its people. Unfortunately they are also open sewers.

America's sick rivers

No water in its natural state is ever one hundred percent pure. Even as it falls from the sky, rain gathers minute quantities of impurities from the atmosphere. Yet living things are remarkably resilient. Plants and animals can tolerate small amounts of impurities in their natural environments and even benefit from them.

We rightly think of man's pollution of waterways as dangerous, even disastrous. But in small amounts, some of the wastes so heedlessly dumped into the water can act as fertilizers. By pouring sewage into rivers, we may supply algae and other plants with vital mineral nutrients that otherwise are in short supply. If not overdone, this can lead to greater productivity of all life in the stream. More often than not, though, we dump too much sewage into rivers; instead of acting like a beneficial "shot in the arm," the injection of sewage turns out to be an overdose that kills the patient. At best, the result is such a profusion of algae that existence becomes impossible for other forms of life. At worst, practically all life in the river is destroyed.

To make matters worse, sewage is only one of the pollutants that are making our rivers sick. Because of careless land-use practices, rivers in many areas are being choked with silt. A fire sweeps through a forest in Minnesota and, after the next storm, torrents of water rush down the hillside instead of seeping into the soil. As a result, a river's health begins to decline. A farmer in Vermont plows furrows up and down a hill instead of across the slope. Rain falls, gullies begin to form, and his river begins to die a slow death. A new highway is slashed through a hill in California, leaving raw cuts exposed to erosion. Another river suffers as a result.

For silt, just as surely as sewage, can kill the life in a river. Erosion is a natural process that has been going on since the beginning of time. But through greed or carelessness or simple ignorance, our mechanized society has increased the natural rate of erosion beyond calculation. Each year millions of tons of topsoil are lost from the land in this way, and thousands of miles of once-clear waterways become rivers of mud. The result does not merely make the rivers unattractive. The murky water blocks out life-giving sunlight; plant growth is slowed; and, by a chain reaction, all animals in the river suffer the consequences as their supplies of both food and oxygen dwindle. Or the choking load of silt may directly affect fish, mussels, and other animals by clogging or injuring their gills so that the animals literally suffocate.

Sadly, much of this senseless waste of topsoil and the resultant damage to our rivers is unnecessary. We know how to prevent erosion, or at least how to slow it down. But too few of us put our knowledge of soil and water conservation into practice.

Contour cultivation of farmland not only assures more bountiful harvests but also helps keep our rivers clear of choking silt. Although such erosion-control methods are well understood, too frequently they are ignored. Both the land and our waterways are damaged as a result.

Kinds of pollution

One form of pollution does not involve any foreign substances at all. The water is simply too hot for animals to survive. Industrial plants often pump vast quantities of water from rivers and use it for cooling purposes. When they dump it back into the river, fishes and other animals gasp momentarily for breath and then float dead upon the surface.

Nothing has been added to the water; it is as "pure" as it was when it was pumped from the river. But many animals have such narrow temperature tolerances that they simply cannot survive in abnormally warm water. Even more important, as we have seen in the case of hot springs, is the fact that the rate of body processes, and therefore the need for oxygen, increases as temperatures rise. In very warm water, a fish's oxygen requirements—even when it is resting—may be greater than the amount available in the oxygen-poor warm water. Thus, one of the most common causes of death in abnormally warm water is suffocation.

There are hundreds of other causes of stream pollution. Lead and zinc waste from mining operations, sulfuric acid, oils, insecticides, household detergents, and other chemicals pour relentlessly into our streams and rivers. Some take their toll by poisoning wildlife immediately. Others accumulate

Clear but lifeless, this stream has been ruined as a habitat for plants and animals. The water is tainted with acids and iron compounds seeping from a nearby strip mine that was simply abandoned when all the coal had been removed. Filling depleted strip mines with soil can prevent this kind of damage.

gradually in the bodies of plants and animals until the build-up becomes fatal, either to the creatures themselves or to other animals that feed on them. Many of these chemicals can be removed from the water, or else treated so that their poisonous effects are reduced. But since each chemical requires a different treatment, it is usually too expensive to deal with all of them at once. As a result, far too many industries ignore their responsibility to other users of the water and make no effort at all to treat wastes of this sort.

But the commonest pollutants are the organic wastes in sewage and in discharges from canneries, paper mills, and other industries. Because the river flows, it usually provides the simplest and cheapest way of disposing of the incredible loads of waste that pour each day from our homes and factories. We know that the river will soon dilute the wastes and carry them around the bend and out of sight. As a result, river after sparkling river has been transformed into a stream of filth.

Yet, strangely enough, rivers and streams are capable of cleansing themselves of organic wastes. Through a remarkable natural process, even heavily polluted rivers can be transformed as if by magic into ribbons of clean, drinkable water, provided we give them the necessary time and space for this process to operate.

Waste oil from a Kentucky well produces a spectacular blaze when it is dumped into a holding pond and burned. Although crude, this method represents a step in the right direction. All too often, waste oil is poured directly into streams, where it covers the surface with a suffocating blanket of slime.

The road to recovery

The key to the purification of most organic wastes is living plants and animals. Think again of the rock you picked up in a rushing headwater stream. The rich slime that covered it consisted of millions of minute bacteria and one-celled animals and plants. Bacteria are able to digest the organic wastes in water by breaking them down into simpler substances. In order to do this, however, they must have oxygen.

If the river is dosed with a reasonable amount of sewage, bacteria soon digest the organic wastes and other living things suffer no damage. But usually too much sewage is dumped, and because of the abundant supply of potential food, the bacteria undergo a population explosion. With so many billions of bacteria living and feeding in the water, all or almost all of the dissolved oxygen is used up. Suddenly the river is unfit for many forms of life, and the purifying process stops or is slowed considerably.

Upstream, where the water is clean and pure, the bottom is alive with mayfly nymphs and caddisfly larvae. Sunfish, bass, and many other animals swarm through the river. A varied natural community lives in harmony with its environment.

Just downstream from the source of pollution, the water is milky white with accumulated organic matter and reeks with foul odors. The most abundant form of life is bacteria, such as *Sphaerotilus*, the so-called sewage fungus, which often coats the bottom with dense feltlike mats of grayish filaments. Aside from bacteria, practically the only forms of life able to survive here are rat-tailed maggots, mosquito larvae, and a few other creatures that can get oxygen from

When a river is polluted with organic wastes, its ability to support life is greatly altered. Upstream from the source of pollution, many different kinds of plants and animals live in the water; downstream, relatively few species can survive. In the heavily polluted septic zone, bacteria and other decomposers consume practically all the oxygen in the water; only creatures that use surface air can live here. In the recovery zone, oxygen becomes more plentiful and a greater variety of animals survives. As the water gradually is purified, the river's original community of plants and animals once again returns.

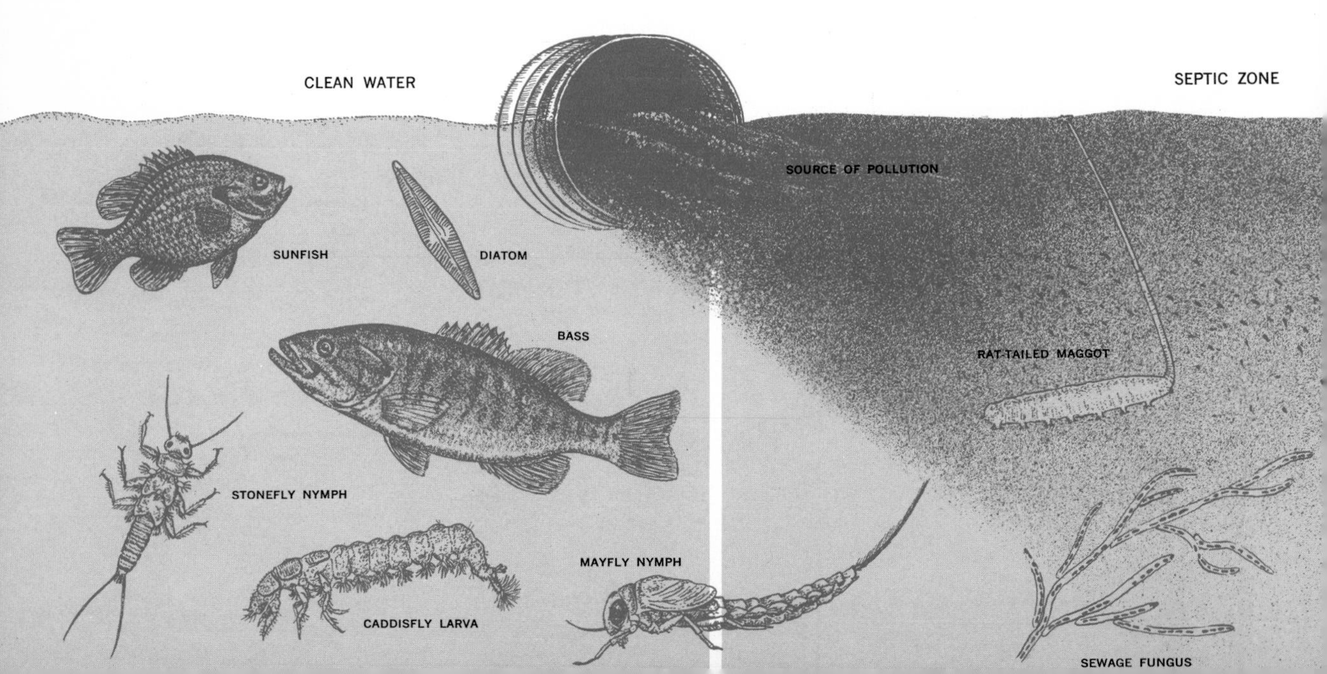

the surface. Any fish unfortunate enough to venture into this area soon suffocates and dies.

Within a few hundred yards, however, living conditions gradually improve as the sewage becomes diluted with water. More algae grow in the river and, in the process of manufacturing food, give off oxygen. More oxygen is absorbed from the surface. The water is still extremely poor in dissolved oxygen, since bacteria continue to consume it, but even this low concentration is enough to support a few more animals.

By now the bottom of the river is likely to be carpeted with masses of *Tubifex* worms, small, slender relatives of earthworms. Half-buried in the bottom muck with their tail ends projecting into the water, great masses of these thread-like worms wave in the current like miniature fields of wind-blown wheat. Like earthworms, they live by eating bits of anything digestible in the threads of mud that pass continually through their bodies. Midge larvae also live in this heavily polluted "septic zone," feeding on bits of waste in the water. Both midge larvae and *Tubifex* worms have red blood. Like our own blood, it contains hemoglobin, which enables them to store oxygen as it becomes available even in small amounts.

Still farther downstream is a "zone of recovery." By the time the water enters this area, some of the organic wastes have been used up. The water becomes clearer. Sunlight penetrates more easily, and still more algae are able to grow. As algae replenish the supply of oxygen, the work of the bacteria is hastened; living conditions improve in the lower recovery zone, and a greater variety of animals begins to appear. The recovery zone grades finally into a zone of clean

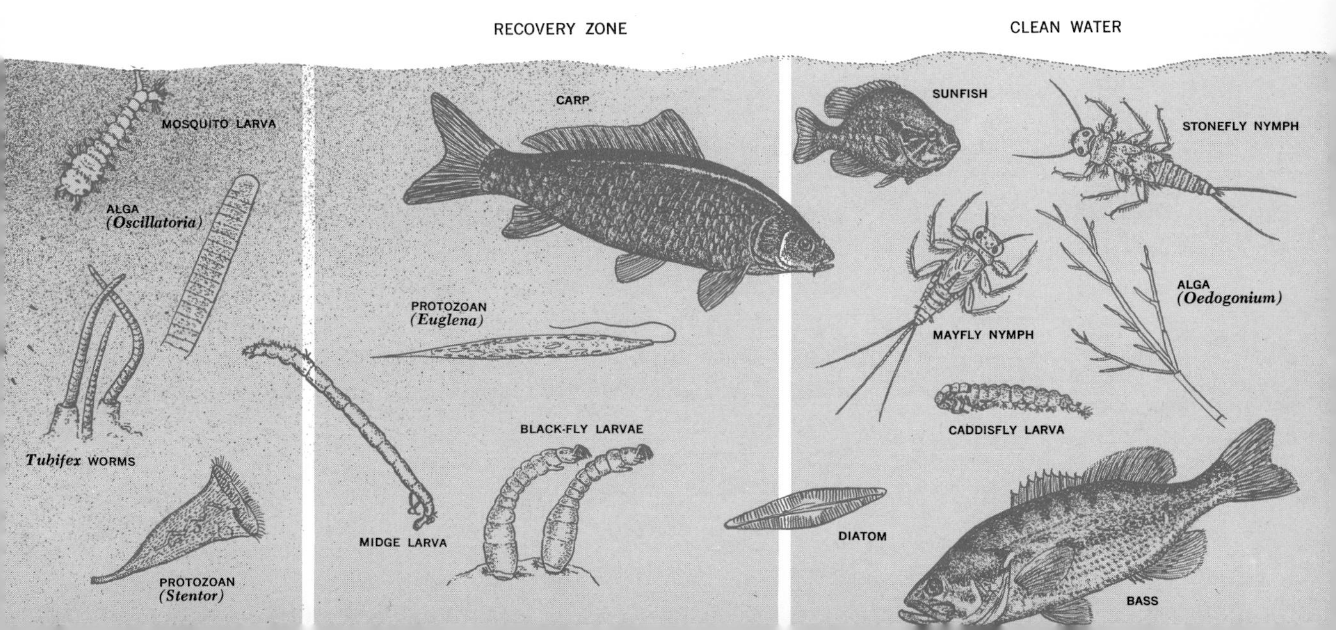

water, where once again there are stonefly nymphs, caddis-fly larvae, bass, and a gradual return of the rest of the river's original community of life.

Even so, streams cannot rid themselves of certain wastes, such as poisons, and of dissolved solids, such as salts. Worse still is the threat posed by disease-producing microorganisms that cause cholera, polio, typhoid fever, and other illnesses. These microorganisms inevitably are dumped into our rivers together with sewage. If the wastes were treated with chemicals in order to kill the disease-causing organisms, the harmless bacteria that purify the river water would be destroyed as well. Thus, the water from most of our inland waterways is seldom safe for drinking unless it is first chlorinated to destroy any disease-causing agents that may be present.

A sewage-treatment plant, shown here in simplified schematic form, closely duplicates the natural process of purification. Just as in rivers and streams, bacteria are the most important agents for decomposing the organic wastes in sewage.

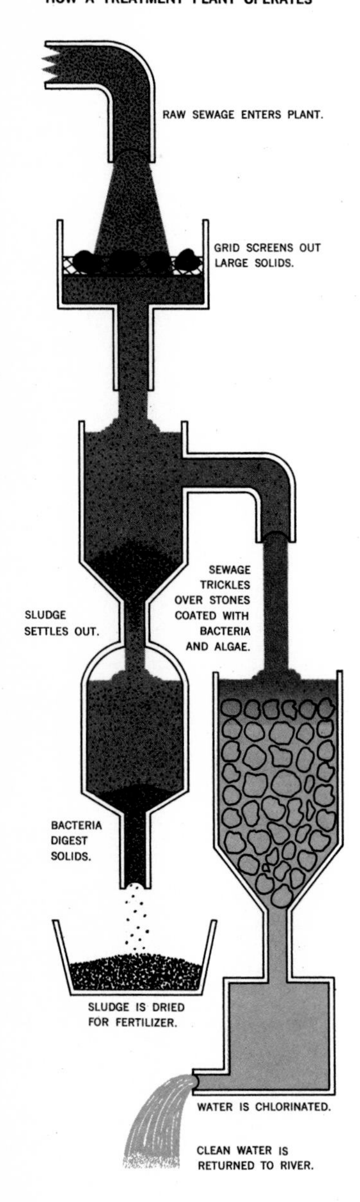

Too much too soon

There are no rules as to how long a stream will take to purify itself. Heavy pollution originating in Chicago has been detected 250 miles away in the sluggish Illinois River. In a swiftly flowing mountain stream, on the other hand, a moderate dose of sewage may be cleared up within 600 yards; the wastes are rapidly diluted, and the oxygen supply is quickly restored.

The rate of purification varies with the size, flow, and temperature of the body of water in relation to the amount of sewage poured in. Unfortunately we usually dump more wastes than the river can handle. Before the water has had a chance to clear, more wastes are added from new sources. Instead of small areas of pollution alternating with long stretches of clean water, far too many of our waterways are polluted throughout practically their entire lengths.

Yet pollution can usually be prevented even in small areas. We know how to process sewage in treatment plants so that the wastes will cause little damage when they enter our rivers. In general, treatment processes duplicate the natural process of purification. Sewage is stored temporarily in ponds or lagoons, for example, while bacteria decompose the wastes. Or sewage is sprinkled over filter beds and trickles down through rocks that become coated with the same slick growth of small organisms that we saw in a natural stream. As yet, though, too many cities, industries,

and individuals ignore their responsibilities for keeping our waterways clean and persist in pouring organic wastes into rivers without treating them at all.

This sewage-treatment plant at Albuquerque, New Mexico, employs the methods shown diagrammatically on the opposite page. If our rivers are ever to be rehabilitated, more treatment plants must be built by communities all across the country. The cost is high—estimates run into billions of dollars—but the need is urgent.

New problems for our rivers

In the past few decades, three entirely new kinds of pollutants have begun to choke our waterways. The deadly three—household detergents, poisonous insecticides, and radioactive wastes—pose especially sinister problems since they defy the usual methods of treatment.

Detergents probably are more a nuisance than a hazard. Since they are not broken down by the usual treatment processes, they remain in water indefinitely. As a result, billowing clouds of foam sometimes drift down certain waterways. Or the detergent may reappear as suds in a glass of drinking water. Although some progress has been made in developing detergents that can be broken down

Less responsive than soaps to natural biological decomposition, synthetic detergents may persist in water and produce billows of foam that transform an idyllic stream into an eyesore. Manufacturers have started to produce new detergents that are more readily broken down, but even the new formulas have not solved the problem completely.

naturally, the problem still has not been solved completely. For example, some of the newer detergents can be broken down quickly, but in the process they produce poisons that are fatal to fish.

Insecticides pose a more serious threat. Many synthetic poisons, such as DDT, have been developed in recent years. Their widespread use results from the fact that they *are* poisonous; they are extremely effective for killing animals that man considers pests. Unfortunately they kill a host of other animals as well.

Every year tons of insecticides are sprayed over marshes to control mosquitoes. Entire forests and vast tracts of farmland also are sprayed to kill insect pests. But all the poisons do not remain in the fields and forests. When rain falls, the insecticides are carried off into rivers and streams. Although by this time the poisons are greatly diluted, scientists have found that even minute traces of insecticides can kill certain aquatic organisms.

Worse still, many of the poisons are very persistent. A fish eats an insect dosed with DDT and, instead of disappearing, the poison is stored in the fish's own body. The more poisoned insects the fish eats, the more insecticide it accumulates in its body. An eagle or grebe then eats a number of poisoned fish and in the process acquires an even larger dose of insecticide. Thus as it moves along a food chain, the concentration of a poison may increase. Eventually it may reach a lethal level and kill a second- or third-order consumer or interfere with its reproduction. Little is known of the possible long-term effects of these potent poisons. A great deal more research is necessary before we can fully understand the hazards involved in using them.

The same is true of radioactive wastes from atomic energy plants on the Columbia and Savannah Rivers. We still are not sure what long-term effects these wastes may have on aquatic life. Even though water may hold only small, supposedly safe amounts of radioactive materials, it is probable that they may be stored in the bodies of living animals. Possibly, like insecticides, the concentrations of radioactive substances increase as they pass along food chains. The end result may be an animal at the last link of the food chain whose body is dangerously radioactive. Although the Atomic Energy Commission has begun studying this unique twentieth-century problem, a great deal more research will be needed before all the mysteries are unraveled.

Can the problems be solved?

Pollution is a many-headed monster. It involves far more than just sewage and silt. In a sense, even using the banks of a wilderness stream for a garbage dump is a form of pollution. By transforming a place of beauty into an ugly blot on the landscape, the dump diminishes the stream's value as a place for recreation.

Yet the problem is not hopeless. In many cases the health of our once-productive streams and rivers can be restored. We know how to reduce the load of organic wastes that clogs so many of our waterways; all we lack is willingness to spend money for modern sewage-treatment plants. We know how to keep our precious topsoil on the land where it belongs; all we lack is determination to put conservation theories into practice. We know how to harness our inland waters and make them work for us in many ways: as transportation channels, as sources of drinking water, as generators of electric power. All we lack is the good sense to treat

CHEVROLET

our streams and rivers with the respect they deserve before we ruin them completely.

For our waterways are something else too. They are rich habitats that abound in wildlife. They are places where we can swim, boat, camp, or simply marvel at the intricate workings of the natural world. Or at least a few of them are.

Rivers for people

In the face of our expanding population, more and more of the wild, unspoiled corners of our land are being destroyed. Mushrooming suburbs, factories, and highways all are taking their toll of open space. Fewer and fewer of our rivers and streams are free of scars inflicted by human carelessness or greed. The time has come to decide whether or not the special beauty of a free-flowing mountain stream or river is worth preserving. Should some of them be set aside as reminders of what all our waterways once looked like?

Fortunately, a few of them are being saved. A number of fine rivers have been spared from destruction simply because they happen to flow through national parks that were established primarily to preserve other features of the landscape. One area in the National Park System, the Ozark National Scenic Riverways, was set aside specifically to preserve portions of two exceptionally beautiful rivers. Here, in the heart of Missouri, 101 miles of the scenic and unspoiled Current River and 39 miles of the Jacks Fork River have been declared permanently off-limits to developers.

Even more exciting are proposals to establish a "National Wild Rivers System," to be administered by the National Park Service and other government agencies. This plan would prohibit the building of dams—or anything else—on a number of rivers so that Americans now and in the future can know the special grandeur of a truly wild, unspoiled river. Parts of six rivers are included in the proposal: the Green River in Wyoming, the Rio Grande in New Mex-

When the banks of a pleasant stream are used as a garbage dump, the scar remains for generations. Running waters usually can rid themselves of organic wastes, but this monument to human abuse probably will mar the landscape for years to come.

OZARK NATIONAL SCENIC RIVERWAYS

Deep in the Ozark Mountains of Missouri, portions of two gemlike rivers, the Current and the Jacks Fork, have been permanently incorporated into our National Park System. Here the vacationer finds an opportunity to savor two American waterways that have been relatively little affected by civilization. Whether he chooses to paddle a canoe (*above left*) or to travel in an outboard-powered "johnboat" (*right*), he can explore a total of 140 miles of natural, unspoiled river. Fishing is excellent, and the scenery is quietly beautiful. In some places, the rivers are bordered by clean gravel beaches for campsites (*below left*). Elsewhere there are tall limestone cliffs punctuated by caves, while the neighboring forests abound in wildlife and Indian relics. Plans are underway to preserve a number of other American rivers in the "wild" state; if these proposals are carried through, we will have created a priceless legacy to pass on to future generations.

ico, the Rogue in Oregon, the Salmon and Middle Fork of the Clearwater in Idaho, and the Suwannee in Georgia and Florida. Eleven other rivers scattered across the country are listed for possible inclusion in the future.

Yet it would be a sad blot on our record as a civilization if, come 1980 or the year 2000, these few rivers were the only clean ones left. Fortunately, a new sense of responsibility and an awareness of our urgent need for clean water seem to be dawning across the continent. Steps already are under way to clean up the Ohio, the Delaware, and many other rivers. Citizens of New York State recently agreed in a landslide vote to spend the billions of dollars necessary for curing the sickness of their rivers. The President has pledged his determination to make the Potomac, which flows beside the nation's capital, a model of beauty and recreational use for the entire country. Only when these ambitious programs have been accomplished will we be able once again to point with pride and say, "*These* are America's great rivers."

Bathers in an idyllic setting at Grand Canyon National Park enjoy one of the finest values of America's rivers. Unfortunately, too many of our waterways have been so badly abused that they are all but useless for recreation. Yet in many cases their beauty and grandeur could be restored—if only we were willing to pay the price.

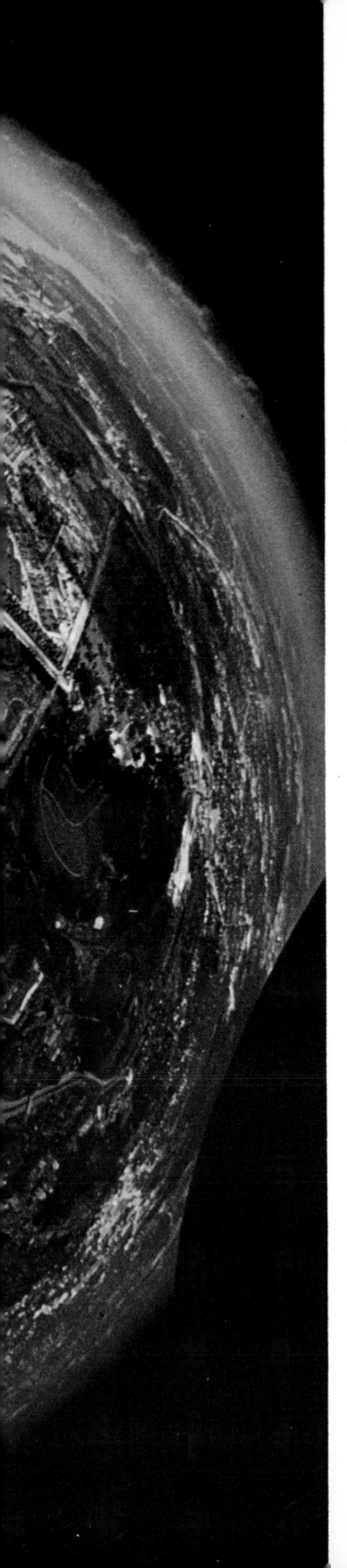

Land of Many Rivers

It is a morning late in April and mist hangs heavy over the water. A fisherman warms up the motor on his small boat. Soon he is heading upriver, shattering the water's glassy surface with a trail of waves. On the opposite bank tall cliffs of solid rock rise abruptly from the very edge of the river. But the fisherman can barely discern their outlines today, for the sun has not yet burned off the gray haze of fog. The only sound that breaks the stillness of the morning is a distant muffled honking, as high overhead a flock of geese heads north to the summer breeding ground near the Arctic Circle.

The fisherman is not on some remote waterway flowing through the wilderness. The river is the Hudson, and just a few miles to the south lies New York City. Factories and tall apartment houses line the river's east bank. To the west, just beyond the crest of the famous Hudson Palisades, other crowded cities stretch on for miles. Yet the fisherman still plies his ancient trade.

Like generations of men who lived in the Hudson Valley before him, he sets out each spring to harvest some of the thousands of shad that swim upriver to spawn. Like their smaller relatives, the alewives, and a number of other fish, shad live in the sea but return to fresh water to lay their eggs. By the end of summer, the young will have grown to four or five inches in length and will return to the ocean. Because of their delicate flavor, thousands of the fish are netted for food as they journey upstream each spring in larger rivers all along the Atlantic Coast and in a few places in California.

Yet when the fisherman hauls in his net, he finds few shad entangled in its mesh. Each year he seems to be less successful than the year before. Regretfully he concludes that next year or the year after he may have to give up fishing for shad altogether. Their spring runs up the Hudson seem to be dying out.

The same thing is happening in rivers all along the coast; where thousands of pounds of shad were harvested in the past, now it is hardly worth the bother of setting and tending the nets. The catch is carefully regulated by law to make certain enough fish get upstream to breed. But in many rivers, the harvest has been too great in the past and shad runs are dwindling. Worse still, the lower Hudson, like many other rivers, is badly polluted with sewage. For several miles, the water is so foul that the fish have difficulty getting past to cleaner water upriver. Well upstream, near Albany, pollution becomes even worse. In some rivers the pollution has been corrected and shad are gradually returning. But so far New Yorkers have barely begun the enormous job of cleaning up their most famous river.

Only an hour's drive from the bustle of Manhattan, New York's Bear Mountain State Park preserves the natural beauty of one stretch of the scenic Hudson Valley. In addition to the enormous task of cleaning up the filth in the river itself, conservationists are battling to prevent destruction of the landscape that forms the river's setting.

The wild Hudson

The Hudson is unusual in several ways. Not particularly large as rivers go—its total length is only about 306 miles—it is famous mostly because of the area it serves. At its mouth, the Hudson flows through one of the most densely populated areas in the country. Yet only thirty or forty miles upstream, its channel slips through a stretch of rugged, wild mountains. This area, known as the Hudson Highlands, is so lovely that many people compare it to the famous valley of the Rhine in Germany. Even now, lawmakers and thoughtful citizens are pondering whether or not industrial development in this part of the valley should be curbed in order to preserve a fine scenic area that lies within reach of millions of people.

The lower half of the Hudson, moreover, is not so much a river as a huge arm of the sea. Much of the bottom lies below sea level, and there are four-and-a-half-foot tides at Albany, 150 miles inland. Salt water extends well upriver, though it lies at the bottom of the channel, with lighter fresh water flowing over it.

Although the lower Hudson Valley is heavily populated and industrialized, this river of contrasts is relatively wild through much of the upper half of its course. From its source at Lake Tear-of-the-Clouds, 4322 feet above sea level, it rushes between rugged forested peaks of the Adirondack Mountains. Here the water is fresh and pure, teeming with trout, dace, suckers, sculpins, and many other fishes. Canoeists may see deer along its banks, as well as raccoons, minks, and a variety of other wildlife. Now and then they may even spot a black bear lumbering off through the forest.

The upper Hudson is so wild and free, in fact, that it is difficult for visitors to realize how sick its waters will become by the time they reach the sea. In this, the Hudson, despite its unique character, is a fairly typical American river: few have managed to escape the mark of man.

Feldspar Brook, spilling from Lake Tear-of-the-Clouds in New York's Adirondack wilderness, is the humble beginning of the majestic Hudson River. Farther downstream on its three-hundred-mile course to the Atlantic Ocean, the river exemplifies both the best and the worst effects of man's impact on our waterways.

Because of the pressures of large populations and heavy industrialization, long stretches of the Hudson River are unfit for human use. Yet its health could be restored by putting into practice the principles of good river management.

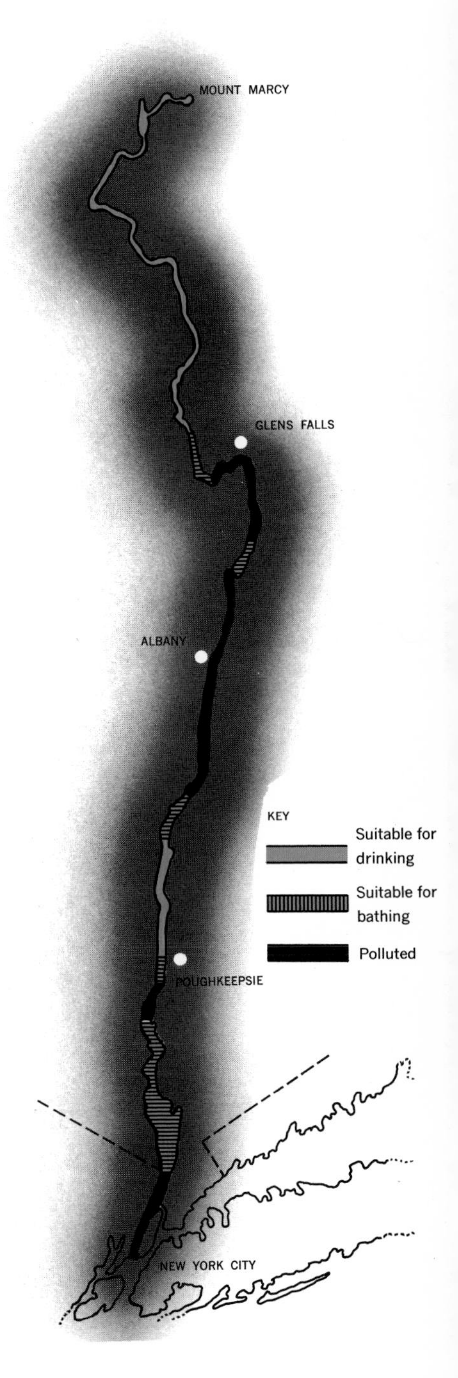

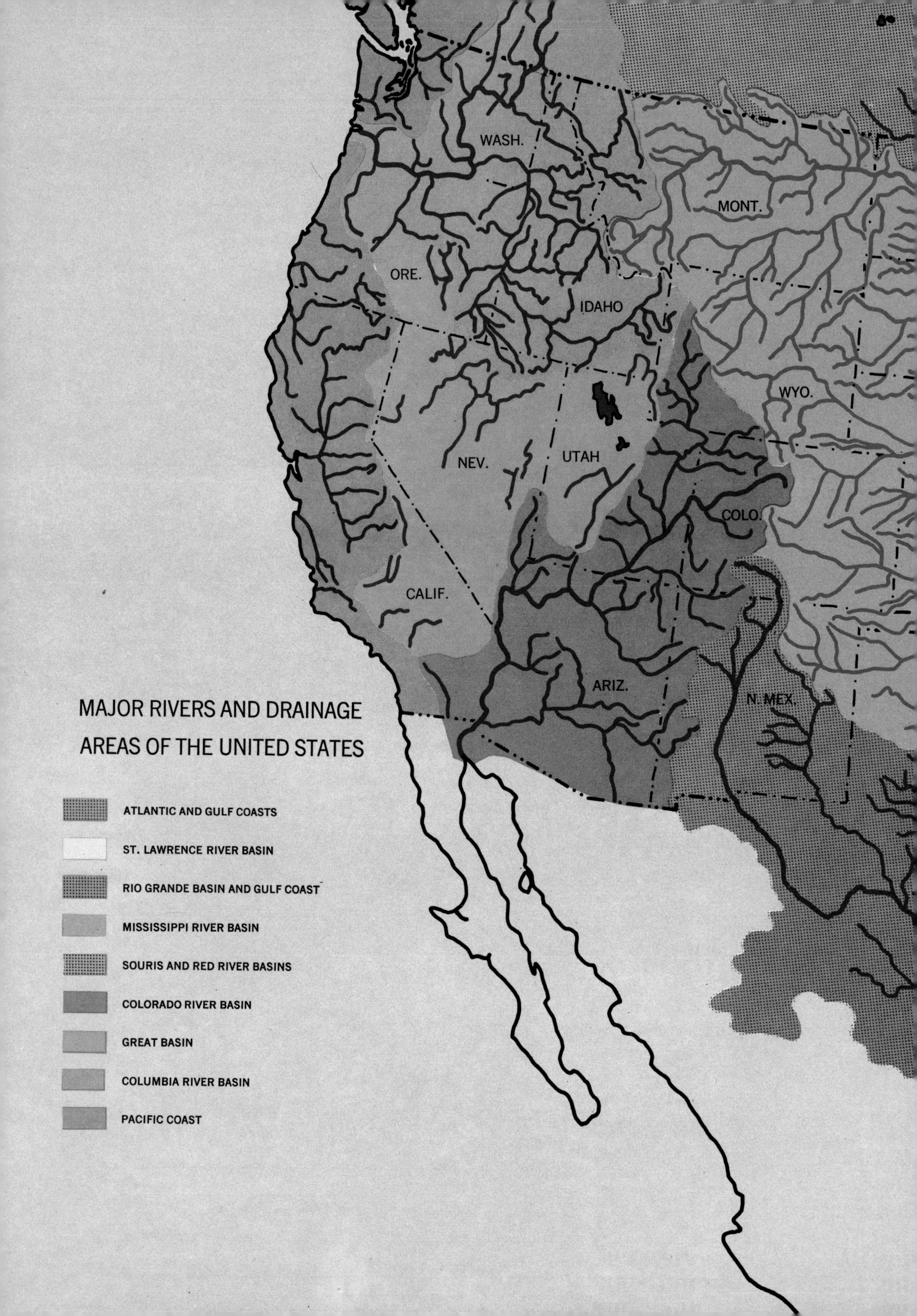
WASH.
MONT.
ORE.
IDAHO
WYO.
NEV.
UTAH
COLO.
CALIF.
ARIZ.
N. MEX.
MAJOR RIVERS AND DRAINAGE
AREAS OF THE UNITED STATES
ATLANTIC AND GULF COASTS
ST. LAWRENCE RIVER BASIN
RIO GRANDE BASIN AND GULF COAST
MISSISSIPPI RIVER BASIN
SOURIS AND RED RIVER BASINS
COLORADO RIVER BASIN
GREAT BASIN
COLUMBIA RIVER BASIN
PACIFIC COAST

MINN.
N. DAK.
S. DAK.
NEBR.
KANS.
OKLA.
TEX.
IOWA
MO.
ARK.
LA.
WIS.
ILL.
MICH.
IND.
OHIO
KY.
TENN.
MISS.
ALA.
GA.
FLA.
S.C.
N.C.
VA.
W.VA.
MD.
DEL.
PA.
N.J.
N.Y.
VT.
N.H.
MAINE
MASS.
CONN.
R.I.

The St. Lawrence

All along the East Coast there is a series of fairly short but well-known rivers. The Delaware, Susquehanna, Potomac, Roanoke, Savannah, and many others are familiar names, for all have played a role in American history. Their drainage basins are relatively small, for they are cut off from the rest of the continent by the Appalachian Mountains. Yet each of them carries a sizable volume of water to the Atlantic since they drain a relatively humid area.

The only truly great river flowing directly into the Atlantic, however, is the St. Lawrence. Flowing north of the eastern mountain barrier, the St. Lawrence carries the runoff from half a million square miles in the interior of North America into the Atlantic.

Instead of the usual maze of branching streams and

Twin-flight Locks at Thorold, Ontario, serve as a three-step staircase for ocean-going vessels passing through the Welland Ship Canal, between Lake Ontario and Lake Erie. These and other locks and canals linking the Great Lakes have transformed Chicago, Duluth, and other inland cities into major world ports.

smaller rivers, the St. Lawrence has as its headwaters the five Great Lakes. Together they make up the largest body of fresh water in the world. Although the lakes are linked by a series of smaller rivers, each with its own name, the entire system is part of the St. Lawrence River. The St. Louis River, which winds through the forests of northern Minnesota and empties into Lake Superior, is the final source of the St. Lawrence.

Nearly two thousand miles away, just north of Quebec's Gaspé Peninsula, the St. Lawrence merges with the sea. At its mouth the river is ninety miles wide, and all the way inland to the city of Quebec it is more like a huge arm of the sea than a river. Between Quebec and Lake Ontario the river narrows considerably. Yet even here, it is not so much a river as a series of locks and ship canals, part of the vast St. Lawrence Seaway. The entire St. Lawrence River, in fact, has been so thoroughly developed that it is one of the busiest waterways in the world. As a result of the locks and canals that connect all the Great Lakes, inland cities such as Milwaukee, Toronto, Chicago, and Detroit have direct connections with the sea. Once a barrier to shipping, even world-famous Niagara Falls on the 35-mile Niagara River, between Lake Erie and Lake Ontario, has been by-passed by the Welland Canal.

Despite all the benefits that have resulted from man-made improvements on this great waterway, however, human alteration of an age-old habitat has caused spectacular damage as well—damage that the canal builders could never have foreseen.

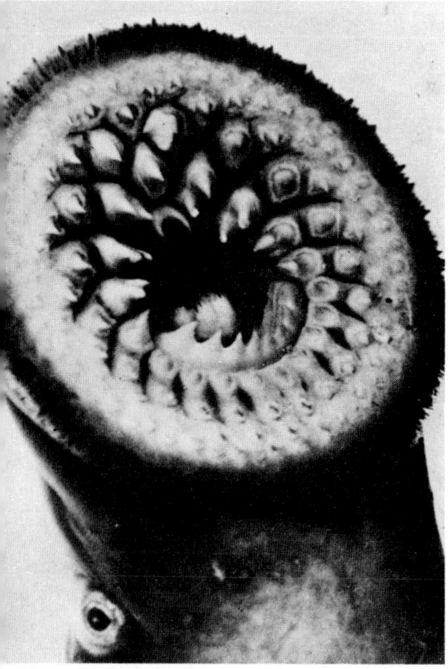

The sea lamprey's mouth is studded with sharp, horny teeth and a rasplike tongue. After attaching to a lake trout or some other fish, the lamprey uses this armament to break through its victim's skin and then sucks its host's body fluids.

Lampreys in the Great Lakes

For untold centuries, sea lampreys—probably landlocked since the age of glaciers—have lived in Lake Ontario and New York's Finger Lakes. Although adult lampreys live as parasites on other fishes, in these lakes they apparently had established a delicate balance with the other fishes and caused little damage.

Lampreys are strange eellike fish, about two feet long. Their skeletons are made up of stiff rods of cartilage rather than true bones. Unlike other fishes, they have no paired fins, and instead of gill covers they have simply a row of small holes on each side of the head. Strangest of all are their mouths, which are not paired jaws, but round disks lined with rows of horny teeth. After attaching their suction-cup mouths to lake trout and other fishes, the lampreys rasp away their hosts' flesh with their teeth and suck the body fluids. In the course of a lifetime, a single adult may kill thirty pounds of fish.

Although the Welland Canal opened the way for lampreys to migrate west from Lake Ontario in the 1800s, no sea lampreys appeared in Lake Erie until about 1920. Like the native species of lampreys already living there, the migrants caused little damage at first. But as they found their way through additional canals to Lake Huron, Lake Michigan, and Lake Superior, the lampreys spread like wildfire. There is no question of the damage they caused. Lake trout, their major victims, had been the mainstay of commercial fishing in the Great Lakes. In the early 1940s, the annual catch of lake trout plummeted from fifteen million pounds to a mere half million pounds. The lake trout could not survive the onslaught of this strange enemy. Fortunately, lampreys in the Great Lakes are now being brought under control and the trout are gradually recovering. But the cost has been tremendous.

After living a year or two as parasitic adults, the lampreys migrate up tributary rivers and streams to spawn. In places with gravel bottoms, they pick up pebbles with their sucking mouths in order to clear out shallow nests. When the female has deposited her eggs in the nest and the male has showered them with sperm, the lampreys cover the nest with gravel. And then they die.

Newly hatched larvae bury themselves in tunnels in the stream bottom and feed on bits of organic material. For four or five years they live this way, causing no harm to other creatures of the stream. But when they are about six inches long, they transform into adults and migrate back to the lakes to take up their parasitic existence.

As a result of years of research, scientists have finally discovered a chemical that, in weak doses, kills the lamprey larvae but does not harm other fishes. One by one, all the streams where lampreys breed in the Great Lakes area are being treated with the poison. Although lampreys are not likely to be eliminated entirely from the lakes, they are gradually being brought under control and a valuable fishing industry is on the road to recovery.

Although sea lampreys are most conspicuous and best known as parasitic adults, they spend the greater part of their life cycle as harmless burrowers in stream bottoms. Control efforts have been concentrated on this stage, and biologists have now developed poisons that kill the larvae in their burrows without harming other fish in the stream.

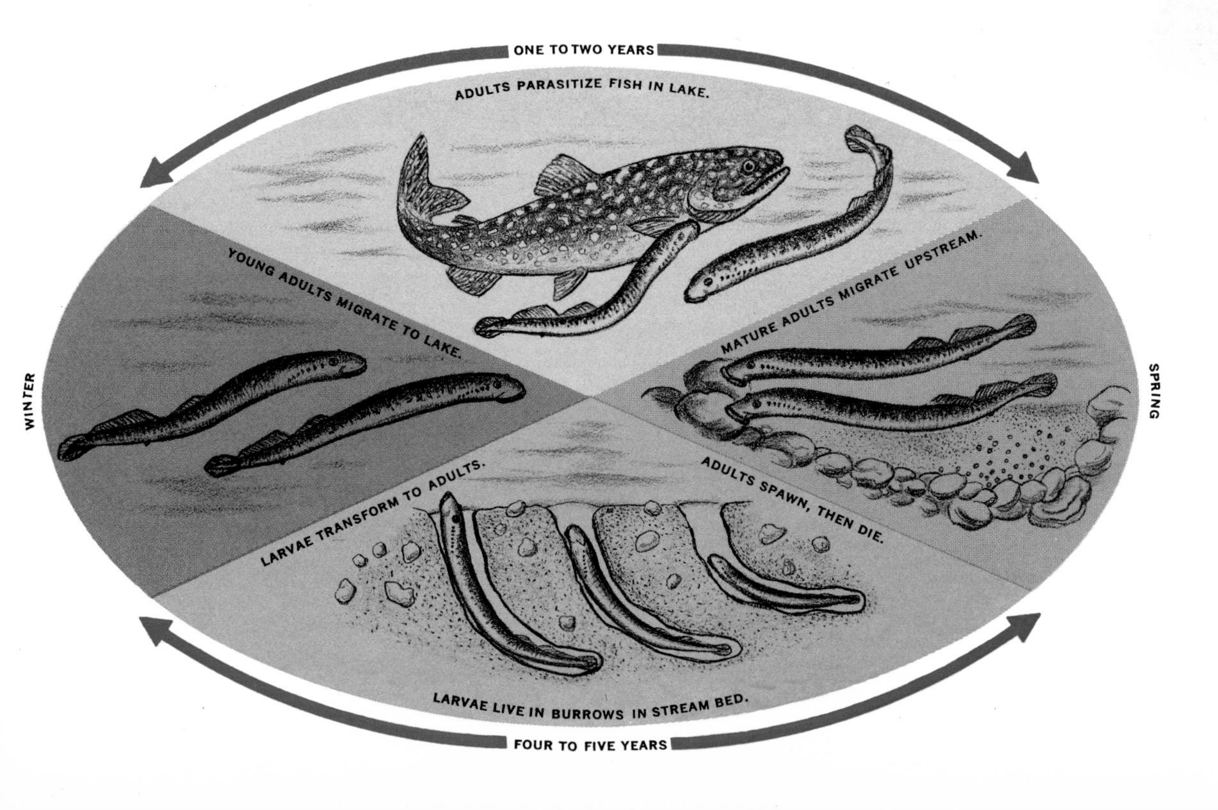

The glow of a winter sunset shimmers across the ice that blankets a portion of the upper Mississippi. When spring thaws finally come and shatter the spell of winter, the river sometimes is clogged for miles by tremendous ice jams.

The father of waters

In northern Minnesota, not many miles from the head of the St. Louis River, the source of the St. Lawrence, lie the small beginnings of yet another great river. Deep in evergreen forests dotted by hundreds of sparkling lakes, there is a relatively small body of water called Lake Itasca. A pleasant brook about ten feet across and shallow enough for wading carries the overflow from the lake down a rocky channel. Each year thousands of vacationers visit the state park that surrounds Lake Itasca, sometimes traveling many miles just to glimpse the outlet stream. For by the time it

winds its way to the Gulf of Mexico, some 2350 miles downstream, this insignificant stream grows to a mile-wide giant: it is the beginning of the Mississippi River.

The river's system of tributaries is incredibly vast. The Mississippi drains the fourth largest area of any river in the world—1,250,000 square miles, or forty-one percent of the United States mainland, excluding Alaska. You can travel from western Pennsylvania to western Montana and from Canada to the Gulf of Mexico without ever leaving the Mississippi River Basin. The drainage area is so huge, in fact, that it is usually thought of in separate units–the Missouri, the Ohio, the Tennessee, the Arkansas, and the Red Rivers.

A tugboat nudges a cluster of barges past one of the tall limestone bluffs that line the Mississippi above St. Louis. Railroads put the steamboats of Mark Twain's day out of business, but with the development of tugboats traffic on the river has become heavier than ever.

The Mississippi's course

As it leaves Lake Itasca, the river wanders through a wilderness of forests and glacial lakes—a remote area where snowshoe hares bound off through the underbrush and the wild call of the loon fills the morning air. Gathering tributaries one by one, it is already a large river by the time it reaches Minneapolis. There, at the Falls of St. Anthony, the Mississippi drops eighty feet within half a mile.

Continuing southward past Wisconsin, Iowa, Illinois, and Missouri, the river glides through rolling prairie country. Tall bluffs rise from the river's edge, and hundreds of islands and shifting sand bars dot its surface. As the Mississippi approaches St. Louis, its water still is fairly clean.

And then it receives a massive dose of muddy water from its longest tributary, the Missouri. For quite some distance downstream, the Mississippi looks like two rivers flowing side by side in the same channel. On one side the water flows relatively clear and blue; on the other side it is thick with reddish silt. Finally the two streams mix, and the Mississippi remains muddy all the way to the Gulf of Mexico.

The Missouri itself is quite a river. From its source in the Rocky Mountains in southwestern Montana, it flows about 2300 miles across the plains and drains all or part of ten states. Some people indeed consider this the true headwater of the Mississippi and claim that the stream flowing from Lake Itasca is simply a tributary of the much longer Mississippi–Missouri River.

The next major tributary is the Ohio River, which joins the main stream at Cairo, Illinois. Formed by the junction of the Allegheny and Monongahela Rivers at Pittsburgh, it flows southwest for 981 miles. Smaller rivers join the Ohio all along the way, but its largest tributary is the Tennessee, which flows into the Ohio River only a few miles upstream from Cairo.

Although the Ohio River is far smaller than the Missouri in the size of its drainage area, it carries a larger volume of water. And although it drains only about one-eighteenth of the land surface of the United States, the Ohio River serves more than one-eighth of the total population.

As it continues south from Cairo to the Gulf of Mexico, the Mississippi is truly a mighty river. Winding beside high

At the turn of the century, paddlefish up to six feet long and one hundred fifty pounds in weight were taken in the Mississippi. Overfishing of these unique river giants and the damming of tributary streams where they spawn have so depleted their numbers that today a paddlefish of any size is a rarity.

levees, its tremendous volume of water slips quietly southward with awesome power. Finally it is joined by the Arkansas and Red Rivers and then, south of New Orleans, flows out in several channels across the swamps of the delta that projects into the Gulf of Mexico.

Floods on the Mississippi

Since the Mississippi funnels the run-off from so large an area into the Gulf, it is no wonder that floods have always been a major problem to people who live along its banks. A heavy snowpack in the Rockies, a sudden thaw in Minnesota, or heavy rains in Ohio could easily make a person homeless though he lived hundreds or even thousands of miles downstream in Louisiana. During the great flood in 1937, for instance, the river rose 56.4 feet at Cairo, Illinois. Almost overnight, 28,000 square miles of the lower valley were transformed into a gigantic muddy lake nearly as big as Lake Superior. When the flood subsided, a million people were left homeless.

Almost overnight, sudden rains or rapid melting of snow upstream can transform the Mississippi Valley into a shallow sea. By constructing dams, levees, and other structures, the Army Corps of Engineers and other governmental agencies are gradually winning the battle to bring this big, unpredictable river under control.

For generations, sandbags have been the basic weapon in the struggle to keep the rampaging Mississippi within bounds. Here workmen are constructing a sandbag battlement to fortify the business district of Dubuque, Iowa, against the rising river.

For many years the only way to combat floods was by building levees, damlike walls that parallel the riverbanks and keep the water in its channel. In many places the channel has also been straightened by elimination of some of the loops. A better long-term solution is the construction of control dams along tributaries to catch heavy run-off upstream and release it gradually into the river. The Missouri River Basin, for example, already includes over a thousand reservoirs, and 150 more are in the planning stages. More dams are also planned in the Ohio River Basin. Besides controlling floods, the dams supply water for irrigation and power for generating electricity.

A better life in the Tennessee Valley

The most dramatic example of overall planning for an entire river basin is in the Tennessee Valley. The Tennessee Valley Authority, better known simply as TVA, was created by Congress in 1933 to develop the entire valley in order to improve the lives of its citizens. Now there are twenty large

The foot-long least bittern *(left)*, smallest of our native herons, is one of many water birds that flourish in the marshes and about the reservoirs of the Tennessee Valley. More likely to be heard than seen, this secretive fisherman makes dovelike cooing sounds as it slinks through dense secluded stands of cattails and rushes.

Of all the migrating waterfowl that pause to rest and feed at impoundments in the Tennessee Valley, few are more attractive than the ring-necked duck *(right)*. It is more easily identified by its brightly patterned bill than by the male's dull chestnut neck ring.

dams on the main stream and on mountain tributaries. Disastrous floods have been practically eliminated in the valley and have been greatly reduced farther downstream on the Ohio and the Mississippi.

But TVA has accomplished more than flood control. As a result of the electric power generated at dams and steam plants, the TVA region now uses about fifty times as much power as it did in 1933. Industry has grown rapidly, and thousands of jobs have been created. By maintaining a navigation channel the full 652 miles from the mouth of the Tennessee to Knoxville, TVA has brought about a thirtyfold increase in river traffic. Soil erosion has been slowed down now that farmers have been taught about conservation methods, and barren hillsides have been planted with trees to hold the soil in place. Malaria has been eradicated and water pollution greatly reduced. The result of all these projects has been a phenomenal increase in prosperity in the valley, where standards of living once were among the lowest in the nation.

As an added benefit, recreational opportunities in the valley have grown tremendously. Sixteen large lakes created by the dams cover 600,000 acres and include 10,000 miles of shoreline. Boat harbors, fishing camps, and summer homes have sprung up around the lakes. Each year, fishermen by the thousands share in the harvest of black bass, walleyes, white crappies, and other fishes that flourish in the lakes. In addition, 100,000 acres have been set aside as refuges for migrating waterfowl. At the peak of fall migrations, as many

The hardy opossum is our only native animal that carries its young in a pouch, kangaroo fashion. Traditionally associated with southern lowlands, it is slowly spreading its range through other parts of the country.

as 200,000 ducks and 10,000 geese pause to rest and feed on Kentucky Reservoir alone. Living conditions in the refuges are so good that many migrating birds remain in the valley through the winter instead of continuing south to the Gulf of Mexico.

The overall objective of the TVA approach to river-basin management is sometimes summed up with the words "multiple use." This means harnessing a river—or any other resource—and developing its potential in a way that produces as many benefits as possible. The concept has worked so well in the Tennessee Valley that, within a single generation, the whole economy and way of life of the people have been completely transformed. Many people feel that other river basins could be improved by developing them in a similar way.

The Rio Grande

Southwest of the Mississippi delta, at the border between Texas and Mexico, another great river empties into the Gulf of Mexico. Besides its obvious importance as an international boundary, the Rio Grande (Spanish words for "big river") is vital to the lives of the people who dwell in this arid region. It is said that only the Ganges in India is more thoroughly used for irrigation and other purposes.

The Rio Grande is born of snow-fed brooks high in the mountains of southwestern Colorado. Near its source it resembles trout streams almost anywhere. Its white water hurtles down a rocky course through mountain forests of spruce, fir, and aspen. In the short distance of eighty miles, it descends some five thousand feet.

Then the slope becomes more gradual and the river flows more serenely through parklike San Luis Valley. There are many ranches here, and the farmers use huge quantities of water from the river and from deep wells for irrigation. So much water is taken from the river that its volume is greatly depleted as it leaves the valley and enters New Mexico.

But the river that flows from San Luis Valley has not always been a weakling. In the thousands of years before men began to rob the Rio Grande of its water, it was a powerful force in molding the landscape. As it enters New Mexico, it flows through Rio Grande Gorge, a spectacular canyon over a thousand feet deep, carved from rock by the once-powerful waters.

Continuing due south through the heart of New Mexico, the river flows through Valle Grande and White Rock Canyon, then past sheer rocky walls where fourteenth-century Indians built cave homes. As it picks up tributaries along the way, the Rio Grande gradually widens and gains new strength. By now it is flowing through semidesert country. Sparse growths of pinyon pines and junipers are mingled

The alchemy of irrigation transforms parched desert land along the Rio Grande into an emerald strip of citrus groves and truck farms. Even before the arrival of Spanish explorers in 1540, local Indians were tapping the "big river" to irrigate their crops. Today a vast complex of dams and reservoirs puts nearly every drop of Rio Grande water to work for man.

with a few prickly-pear cactuses, although many large trees grow near the river. Mule deer, bobcats, coyotes, and even a few mountain lions roam the arid countryside.

On past Albuquerque the river picks up a few large tributaries, plus many smaller streams that are dry most of the time. After infrequent cloudbursts these channels, or dry washes, suddenly seethe with torrents of water. Within a few hours or days the flow slows down to a trickle and the channels finally dry up again—but not until a good deal of water and silt have been dumped into the river.

In southern New Mexico, the Rio Grande suddenly widens into its principal reservoir, Elephant Butte. The lake, over forty miles long, is now a popular recreation area, with fine fishing for bass, catfish, and crappies. Erosion upstream is so severe, however, that scientists estimate that the lake will be completely filled with topsoil in less than a hundred years.

As the river continues southeast and becomes the Texas–Mexico border, it is subject to great variations in flow. So much water is removed for irrigation that, by the end of each growing season, the reservoirs along the way are almost empty and the river becomes a feeble trickle. Legal agreements between various states and Mexico establish the amount of river water to be used by each.

For many miles now, the river continues at a low ebb until it picks up water from the Mexican Rio Conchos. With its volume—and its grandeur—restored, the Rio Grande flows along an enormous curve, passing through a region of spectacular scenery and immense cattle ranches. Big Bend National Park lies at the heart of this wild country and includes three sheer-walled canyons carved by the river. The landscape in the park is unbelievably varied. Cool evergreen forests on the Chisos Mountains give way on the lower slopes to desert vegetation. Mesquite, yuccas, and many kinds of cactuses grow here, and wildlife ranges from jack rabbits and lizards to mountain lions, peccaries, and mule deer. Yet visitors to the fertile bottom lands beside the river can camp in the shade of giant cottonwoods, willows, and other large trees.

Still seen occasionally in the Rio Grande country and elsewhere in the Southwest, the magnificent mountain lion once ranged through most of North America. Relentless persecution by hunters, stockmen, and even government agencies has led to virtual extinction of the big cat over most of its former range.

BIG BEND NATIONAL PARK

Embraced by a hundred-mile-long U-turn of Texas's Rio Grande, Big Bend National Park preserves some of the most unspoiled and scenic river wilderness to be found anywhere in the country. Stretches of sandy lowlands, dense reed jungles, narrow fifteen-hundred-foot-deep canyons—Big Bend has them all. Wildlife also is abundant and varied. A thousand different kinds of plants, including many bizarre desert forms, have been identified within the park. Two hundred species of birds have been recorded. As for mammals, Big Bend offers an opportunity to see ringtails, peccaries, kit foxes, pronghorns, mountain lions, and dozens of other uncommon creatures in their natural habitat.

Leaving the Big Bend country, the river is swelled still further by its largest tributaries. The Pecos, the Devils River, and several that rise in Mexico have joined forces with the Rio Grande by the time it reaches its delta, one of the most unusual areas in the country. The part of the delta that lies in the United States is about seventy miles long and thirty miles wide.

The delta is known locally as the "magic valley." Until about twenty-five years ago, it was covered by a thorn forest of mesquite, cactuses, and a host of unusual trees and plants found nowhere else in this country. But now bulldozers have changed the delta almost beyond recognition as the land has been brought under cultivation. Since it is nearly frost-free, this incredibly rich agricultural area supplies a major portion of winter vegetables for the rest of the United States.

Irrigation needs for all this farming are tremendous. As it flows across the delta, almost all the water in the Rio Grande is used for agriculture. Relatively little is "wasted" by flowing into the Gulf of Mexico. Thus eighteen hundred miles from its small beginnings in the mountains, the Rio Grande in a sense is still a small river. All along its course, water is removed almost as quickly as it flows in, as the Rio Grande works to satisfy the endless thirst of man in an arid region.

The Continental Divide, the natural "backbone" of North America, separates all our rivers and streams into two great families. Those to the west of the divide flow into the Pacific Ocean, the Gulf of California, or the Great Basin. Those to the east flow into the Atlantic Ocean or the Gulf of Mexico.

Across the Rockies

If you should ever drive through the Rocky Mountains, perhaps on a tour of the West's national parks, time and again you will pass highway signs marking the Continental Divide. The signs invariably stand at the tops of passes through the mountains. They indicate the division line between the two great watersheds of North America. East of the divide, all the water running off the land from rain and melting snow flows eventually into the Atlantic Ocean or the Gulf of Mexico. Water rushing down the western slopes of the mountains drains into the Great Basin, the Pacific Ocean, or the Gulf of California.

If you examine a map of the United States, you will probably find a jagged line representing the Continental Divide. From western Montana, the line angles southeast to central Colorado, then wanders south through western New

Mexico before passing into Mexico. Like a seashore, the Continental Divide is a great natural boundary, not a man-made one such as a state line. Every watershed, of course, is surrounded by its own divide, a fringe of high land marking the limits of the area it drains. But the Continental Divide is the biggest and most important one in North America. Rivers flowing west of the divide include the Columbia in the North, the Sacramento and San Joaquin in the Central Valley of California, the Colorado in the Southwest, and a number of smaller rivers.

The Colorado River

Like the Rio Grande, the Colorado rises in the Rocky Mountains. Its source is meltwater from heavy winter snows on the high peaks of northwestern Colorado. Almost alone, these snowfields keep the river flowing for 1450 miles through arid plateau and desert country.

Over much of the Colorado River Basin, rainfall is less than fifteen inches a year, and it amounts to a mere five inches in the extreme South. In contrast to the green, well-watered valley of the Tennessee and other eastern rivers, the Colorado flows through a region where life is sparse. Instead of gently rounded hills covered by forests and farmland, there are barren plateaus, sheer cliffs, and vast deserts where plants are few and far between. Here, even more than in the

Bighorns, or mountain sheep, are a thrilling discovery for hikers near the rugged Rocky Mountain headwaters of the Colorado River. The sheep in this group are immature. When fully grown, each one will wear a pair of wide spiraling horns weighing as much as forty pounds.

FLASH FLOOD!

Here today, gone tomorrow, Grand Falls on the Little Colorado River in Arizona are spectacular, if fleeting, testimony to the brutal power of a desert flash flood. Within hours, a heavy rainstorm upstream can fill the nearly dry riverbed with a rampaging torrent that transforms the cliffs at Grand Falls into a thundering cataract of rock, sand, and chocolate-colored water. At its peak, the fury of the river is unimaginable: hundred-ton boulders are plucked from the riverbed and hurled like pebbles down the 185-foot precipice at the falls. Yet only a few hours or at most a day or two later, the fury of the flood is spent and the river subsides again to a feeble trickle.

HARNESSING THE COLORADO

Like a gigantic concrete plug, Hoover Dam *(left)* blocks a canyon on the Colorado River between Arizona and Nevada. This impressive feat of engineering—the dam is 1244 feet long and 726 feet high—is an example of a wild river tamed and put to work for man. Before the dam's construction in the early 1930s, the Colorado River each year scoured large areas of the Southwest with rampaging floods fed by melting mountain snows. The floods often were followed by equally devastating droughts, with the river drying up altogether.

Hoover Dam, Davis Dam, and a series of others have broken this cycle of alternating flood and drought. Spring torrents now are stored in Lake Mead and other huge reservoirs behind the dams, and later in the season the water is gradually released as it is needed for irrigation. In addition, water spilling over Hoover Dam spins a battery of immense generators *(right)*, providing a bounty of low-cost electric power for the Southwest. Lake Mead itself—over one hundred miles long and up to eight miles wide—is an enormous oasis in the desert, providing unequaled recreational opportunities for millions of Southwesterners.

valley of the Rio Grande, the lack of water limits the growth of both human and wildlife populations.

As a result of man's efforts to harness the river for human needs, the Colorado has been greatly altered over time. In some places, traces still remain of irrigation canals that were built by Indians long before the arrival of the first settlers. But the greatest changes have come in the last century. Over a dozen major dams have been built in the Colorado River Basin, and at least twenty-four more are planned. Among the most famous are Glen Canyon Dam and its reservoir, Lake Powell; Hoover Dam and Lake Mead; Davis Dam and Lake Mojave; Parker Dam and Havasu Lake; and the Imperial Dam and its reservoir.

Besides contributing to flood control and providing electric power and water for irrigation and human consumption, the giant reservoirs are a welcome source of recreation in the Southwest. Lake Mead National Recreation Area, for example, is visited by tens of thousands of people every year. Fishing, boating, and swimming all have their followers, but many people come just to admire the surrounding desert and its great stands of cactuses and Joshua trees. Some hope to glimpse a mountain sheep, or "bighorn," or even a mountain lion. Still others come to watch the throngs of ducks, geese, herons, egrets, and other water birds that are attracted to this oasis in the desert.

The reservoirs are in danger, however, for they are rapidly filling with silt. Even the early explorers found a river that was muddy red from its heavy load of silt. In the past, scientists have estimated that as many as a million tons of silt and sand passed by one point in the river every twenty-four hours. Now the river runs fairly clear except after heavy rains. As its flow is interrupted by one dam after another, most of its load of soil settles out in the reservoirs.

The dramatic beauty of the desert landscape is largely the result of natural erosion. Rain is infrequent in the Southwest, yet storms are sudden and violent. For thousands of years, the rapid run-off after each storm has carved the land mercilessly and carried its topsoil to the sea. The

MAJOR DAMS IN THE COLORADO RIVER BASIN

This magnificent formation, the "Cathedral in the Desert," has been lost forever beneath the reservoir behind Glen Canyon Dam on the Colorado River. Is such destruction too high a price to pay for the benefits derived from dams? The question is highly controversial.

river, in fact, probably is best known as an erosive force. Halfway down the Colorado's course to the Gulf of California, its swiftly flowing waters rush through the world-famous Grand Canyon. This colossal gash in the earth was carved over thousands of years by the load of rocks, boulders, pebbles, and sand grains that are hurtled downstream by the powerful current.

The canyon, which is now preserved in a national park, is a mile deep and, on the average, ten miles wide. It is so immense that from its rim the river looks like a slender ribbon of silvery water, seeming far too frail to have carved such a wound through solid rock. But those who descend to the bottom discover that the Colorado is a mighty river indeed. Over a hundred yards across and twelve to forty feet deep, the river rushes downstream with speeds up to twenty miles an hour. Here, more than anywhere else in the world, a visitor can sense the true impact of erosion. To stand beside the river and gaze up the towering canyon walls is to grasp once and for all the awesome power of running water.

North to the Columbia

Traveling north along the Pacific Coast, a visitor would cross many rivers, one after another, like climbing the rungs of a ladder. Most of them are fairly small, for they rise in the mountain ranges that closely parallel the coastline. The only large ones are the Sacramento and San Joaquin Rivers, which merge and flow as a single river into San Francisco Bay. These two rise farther inland and drain the great Central Valley of California between the Sierra Nevada and the Coast Range.

But the largest river on the Pacific Coast is the Columbia. Near the sea, it forms the boundary between Oregon and Washington. Its source lies twelve hundred miles upstream in British Columbia, just north of Idaho. Its main tributary, the Snake River, originates in Yellowstone National Park, Wyoming, and joins the Columbia in southwestern Washington. Between the two, they drain about a quarter of a million square miles of land.

Much of the Columbia River Basin still is fairly wild country, but the potential wealth that could result from carefully planned development is incalculable. Major dams

The mountain beaver is a resident of dense forests found in parts of the Columbia River Valley and elsewhere in the Pacific Northwest. Its name is misleading, since the chunky cat-sized mammal is neither restricted to mountains nor closely related to beavers, beyond the fact that both are rodents. An expert burrower, it builds elaborate networks of underground tunnels, but it comes regularly to the surface to gather ferns, grasses, twigs, and other plant material which it stores for use as winter food and bedding material.

Two hundred and seventeen miles long, four to eighteen miles wide, and one mile deep, the Grand Canyon of the Colorado River in northwestern Arizona is an awesome tribute to the erosive powers of rain, wind, and running water. Seven million years ago—a rather short time, as geological intervals go—the site now occupied by the canyon was a wide level plain traversed by a river that probably looked relatively placid. But then, as a result of a shifting of the earth's crust, the plain began to arch upward into a dome. As the land gradually rose higher and higher, the river cut constantly downward, carving a narrow, ever-deepening scar in the earth. Other natural forces also contributed to the process, and meanwhile wind and frost etched the rocks into strange but strikingly beautiful formations. The result is today's Grand Canyon, one of the most unforgettable spectacles in all of nature.

The view from the bottom of the Grand Canyon is nearly as spectacular as the view from its rim. The sand, heaped into gigantic dunes by the wind, was left behind when the river receded after a seasonal flood.

The angry waters of Granite Falls Rapid are impressive testimony to the grinding power of the Colorado River. Until recently, a million tons of gravel and other abrasives were swept past any given point on the river in a typical twenty-four-hour period, but the construction of Glen Canyon Dam upstream from the canyon has somewhat subdued the rampaging Colorado.

GRAND CANYON NATIONAL PARK

The scenic beauty of the Grand Canyon has been preserved in one of America's most popular national parks. The reservation includes something for nearly everyone: camping, fishing, hiking, muleback trips down into the canyon itself—or just the chance to gaze at some of the world's most spectacular scenery. For the adventurous, the Colorado River and its tributary streams also provide an opportunity to explore one of the wildest waterways to be found anywhere in North America and to sense the fury of the natural forces that are still at work in the canyon.

on the river so far include several giants, such as Bonneville and Grand Coulee Dams, plus a number of smaller ones. Proper irrigation with the waters of the Columbia could bring four million acres of new land under cultivation. Of the undeveloped potential for generating electric power in the United States, forty percent lies in the Columbia River Basin.

Beautiful waterfalls, spectacular mountain scenery, and thousands of acres of towering evergreen forests that grow along its banks make the Columbia a majestic river. The giant of the western rivers also has fish to match its size. The most impressive is the white sturgeon, an ocean-going fish that returns to the river to spawn. The sturgeon, which has relatives living in the Mississippi and a few East Coast rivers, grows to ten feet in length and sometimes weighs over a thousand pounds. Unfortunately, its eggs are valued as caviar, and sturgeon have been slaughtered so recklessly in the past that relatively few survive in the Columbia or any other North American river. But the most famous fish of the Columbia, and by far the most valuable, is the salmon.

Each year, mature salmon of four species forsake their rich feeding areas in the Pacific and migrate up West Coast rivers to spawn. The migrants often are so incredibly numerous that people travel miles just to watch this dazzling spectacle.

Perched on rickety platforms that jut out over Celilo Falls on the Columbia River, Indians use long-handled dip nets to snatch migrating chinook salmon from the seething torrent. Although an age-old treaty gave the Indians exclusive fishing rights at this spot, the falls have now been inundated by construction of the Dalles Dam.

River of fish

For uncounted generations, Indians have congregated each year along the Columbia and other rivers of the Northwest. With dip nets and spears, they harvest a portion of the throngs of salmon that journey upstream from the sea to spawn. Now the salmon are the basis of an industry that yields ten million dollars or more each year and employs as many as 25,000 people.

Nor are men the only ones who capture salmon. During the spawning season, bears throng to the riverbanks and gorge themselves on fish, especially in Alaska. The bears often plunge directly into the water, even in swift rapids, and snatch the salmon as they make their way upstream. Gulls and eagles usually hover nearby, anxious to share the left-overs when the bears have finished eating.

Of the four kinds of salmon that are netted on the Columbia and sent to canneries, the chinook is the largest. This river giant averages twenty-two pounds, but occasionally a ninety-pounder is caught. The other three—coho, sockeye, and chum salmon—usually weigh between four and ten pounds. One after another, different species begin to enter the river in spring, and spawning runs continue until late fall.

The salmon's life begins in the cold shallow water of tributary streams, sometimes hundreds of miles inland. A few months after they hatch, the young begin a leisurely journey downstream. Feeding on microscopic plants and animals, they grow rapidly and are four to five inches long by the time they reach the mouth of the river.

When they enter the sea, their diet gradually changes and they begin to feed on herring and other small fishes. For the next four or five years, the salmon wander hundreds of miles offshore in the North Pacific, intent only on eating and escaping from predators such as seals, sharks, and sea birds. As they become sexually mature, their habits and appearance undergo startling changes. The jaws of the males elongate and develop ferocious-looking hooks at the end. The silvery coho and sockeye males gradually become brilliant red. Before long, their oceanic wandering comes to an end. They head toward shore and seek the mouths of their ancestral rivers.

Once they enter fresh water, the salmon feed no more. For the next several weeks or even months, until their breeding cycle is completed, they live only on the reserves of fat stored in their plump bodies. As the great schools of fish move upstream, neither rapids nor waterfalls, seething torrents nor swirling whirlpools can stop them. The salmon are compelled by an irresistible urge to beat their way to their breeding waters. If they come to a waterfall, they leap as high as eight or ten feet to cross it. If they cannot make it over the waterfall on the first try, they leap again and again, until they either cross the barrier or fall exhausted at its base.

But lately the rivers have been blocked by barriers the fish cannot cross. As stream after stream is sealed off by high dams, many ancestral breeding waters are being lost to the salmon. Frequently "fish ladders" are constructed beside dams. These are series of pools that enable the fish to make their way, step by step, around the concrete barriers.

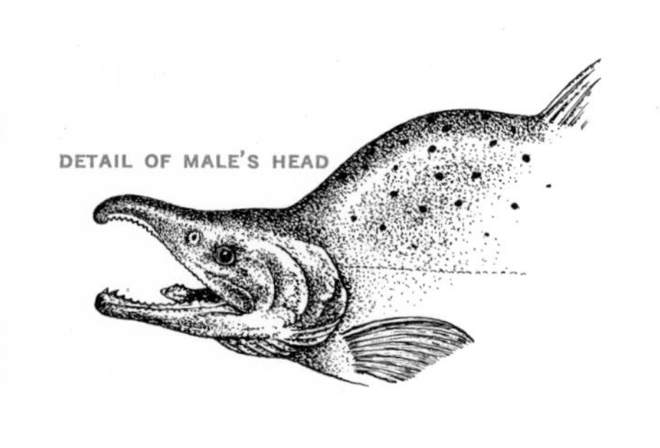

During their spawning migrations, male salmon are easily distinguished from females by their hooked snouts and conspicuously humped backs. These body changes are more obvious in some species than in others, but their significance is not fully understood.

An Alaska brown bear, one of nature's ablest salmon fishermen, scans the water for a possible catch. Standing about four feet tall at the shoulder and weighing as much as three-quarters of a ton, the brown bear is the largest land-dwelling carnivore on earth.

Salmon season is a good time for Alaska's brown bears. In July, as the big fish begin their spawning migrations, bears gather along Alaskan streams for a summer-long feast. Each bear claims its own section of the stream and defends its territory against intruders. Fishing technique varies with different individuals. Some bears remain on shore and snatch passing fish with a lightning-fast flick of the paw; others plunge into the water and catch salmon in their mouths. The cubs are too young to fend for themselves during their first summer, but the following year their mothers lead them to the stream and instruct them in the art of fishing. Perhaps "soloing" for the first time, three bear cubs (*below*) wade gingerly into the shallows of a salmon stream and prepare to try their luck. Although out of the picture, their mother probably is fishing nearby.

Got it! After a series of frustrating near misses, one of the cubs comes up with a salmon squirming between its teeth (*above*). Taking no chances on losing its prize, the cub carries the flopping fish to the safety of higher ground before settling down to a feast (*right*).

But many salmon never find the ladders, and some of those that make it past the dams lose their way in the still water of the reservoirs upstream.

In spite of dams, areas of polluted water, and the relentless harvest of fish by men, birds, and even bears, some of the salmon miraculously complete the journey to their spawning areas. Strangely, experiments with marked fish indicate that each salmon finally makes its way to the stream from which it began its journey to the sea several years before. Although a fish may pass many suitable streams as it travels upriver, it does not leave the parade of migrating salmon until it reaches its own "home" stream.

How do they find their way? Apparently the salmon literally smell their way to the proper stream. Because of the kind of land it crosses and the mixture of plants that grow in it, the water in each stream seems to have a distinctive odor. The odor is too subtle for humans to detect, but evidently it is strong enough to provide the necessary clue for salmon.

In one experiment, several fish were captured from their spawning stream, then released downriver from its mouth. Half the fish had their nostrils plugged; the other half did not. As they swam upstream once again, the fish with unplugged nostrils turned off when they came to the spawning stream. The fish with plugged nostrils could not smell the water flowing from the side stream and simply swam on past.

When the salmon finally reach their destination, they scrape a shallow nest in gravel on the stream bottom. The female scatters several thousand red pea-sized eggs in the hollow, and the male showers them with milt. Their mission now has been accomplished: their journey up the river at last is ended, and the fish have spawned. Scarred and emaciated, the adults drift listlessly downstream, and soon they die.

But next year and the year after, new generations of salmon will return from the sea to spawn and die in the streams where they were born. For thousands of years the

A fish ladder is an ascending series of pools built to help salmon and other fishes by-pass the man-made dams that block many of their migration routes. By permitting the fishes to advance in short leaps, the ladder makes it possible for them to surmount even very high dams.

At the end of a day's fishing, mother and cubs head for home. When salmon are scarce, Alaska brown bears are content to feed on berries, grasses, mice, ground squirrels, and a variety of other foods.

fate of the salmon has been bound to the endlessly flowing rivers. Like hundreds of other wild creatures that live in running water, they will continue to flourish only so long as their rivers flow unpolluted and reasonably free of man-made obstructions.

The beckoning rivers

In this brief tour of some of the major rivers and drainage basins of North America, we have seen how they vary in different areas of the country. Despite their many similarities, each river is unique; each has a personality of its own, as distinctive as the landscape through which it flows.

This is true of every river and stream, large or small, that traces a pattern across the face of North America. Yet the special quality of each is likely to go unnoticed by those who merely glance at the river in passing. The fascination of life in flowing water and the interrelations of the plants and animals that dwell in it are revealed only to those who take the trouble to visit the river themselves and try to understand it.

In this sense, every river and stream in the country, from the smallest mountain trickle to the mighty Mississippi, extends an open invitation. Come to the river and observe its wonders firsthand. Your reward will be the thrill of discovery and the satisfaction that comes from understanding the marvels of the living world of nature.

Each of America's thousands of streams and rivers is a dynamic community of living things. Each one extends to all of us an invitation to pause, to investigate its workings, and to discover the wonders of life in flowing water.

Appendix

Rivers and Streams in the National Park System

The people of the United States are fortunate in owning an extensive system of national parks, monuments, and recreation areas, for these areas encompass some of the finest remnants of the primeval North American wilderness. Only one, the Ozark National Scenic Riverways, in Missouri, was set aside primarily to preserve the natural beauty of a river. Yet there is scarcely an area in the National Park System whose value and interest are not enhanced by streams, rivers, waterfalls, springs, glaciers, or other evidences of water on the move. Without the Colorado River, for example, there would have been no Grand Canyon. And the Rio Grande in Texas both forms the southern boundary and ranks as one of the major recreational attractions of Big Bend National Park.

River-oriented activities in the parks are numerous and varied. The visitor in search of adventure can journey on foot or by mule train to the bottom of the Grand Canyon; in the process, he will enrich his understanding of the erosive power of running water. He can explore rivers of ice, the glaciers on Mount McKinley and Mount Rainier, or visit the underground streams at Mammoth Cave. If his hobby is angling, then he will find unexcelled opportunities for fishing throughout the Park System. (Angling regulations and information about licenses and catch limits, if any, are available at park headquarters in each area.) At the Ozark National Scenic Riverways, canoe trips and wilderness camping are favorite pastimes. In short, the rivers and streams, large and small, that flow through our national parks and monuments provide a valuable bonus of adventure and enlightenment for nearly every vacationer, whatever his interests. Outstanding rivers and other attractions at several areas in the National Park System are described here.

BLACK COTTONWOOD

Big Bend National Park (Texas)

A great curve of the Rio Grande forms the 107-mile-long southern boundary of this park on the Mexican border. The river flows through sandy flats, dense reed jungles, and three 1500-foot-deep canyons carved by mud- and sand-laden waters. The catfish run big, and visitors are free to catch them without a license. Drooping juniper, a tree common in Mexico, is at the northernmost limit of its range in the park. Among the two hundred species of birds you may catch a glimpse of the rare Colima warbler; the park is its only known nesting place north of the Rio Grande. (See pages 174 and 175.)

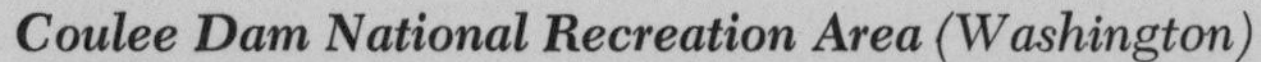

Coulee Dam National Recreation Area *(Washington)*

The focal point of this area is the immense lake behind Grand Coulee Dam, but the area also provides an excellent opportunity for studying the long-range effects of glaciers and water erosion. Glacial meltwater once plunged over a four-hundred-foot rock wall now known as Dry Falls, and near Banks Lake you can walk on land that, in the age of the glaciers, was the bottom of a river channel. Canada geese, brant, mallards, canvasbacks, redheads, and wood ducks are found in the area. The fish are trout—Dolly Varden, rainbow, and Kamloops, a hatchery-developed breed—and bass, pike, sunfish, and crappies. A scenic road runs along the banks of the winding San Poil River.

REDHEAD

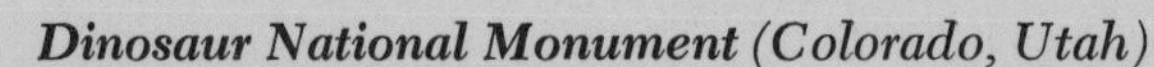

Dinosaur National Monument *(Colorado, Utah)*

The Green and Yampa Rivers have carved spectacular canyons in this monument. At Dinosaur Quarry, where fossil bones have been found in the rocks, you stand on a spot believed to have been a sand bar in a shifting river channel millions of years ago. Before the mountains rose, the whole area was a plain with sluggish streams twisting through mud and sand. Today it is semidesert country populated by turkey vultures, eagles and other hawks, and a variety of smaller birds. There are catfish in the rivers and rainbow trout in the Jones Hole Creek, a tributary of the Green River.

Glacier National Park *(Montana)*

Glaciers and streams shaped the landscape of this wilderness area straddling the Continental Divide. In addition to gemlike lakes, spectacular valleys, and natural amphitheaters gouged from mountainsides by ancient glaciers, the park contains a number of smaller, still active glaciers. The lakes and streams harbor pygmy whitefish, found only in a few places in North America; kokanee, landlocked sockeye salmon; grayling; pike; and cutthroat, rainbow, and Dolly Varden trout. Black cottonwood, quaking aspen, alder, willow, and black hawthorn grow along the streams. Among the birds are the dipper, osprey, bald eagle, and the elusive pileated woodpecker. The park's most famous mammal is the bighorn, or mountain sheep.

Glen Canyon National Recreation Area *(Arizona, Utah)*

The major attraction of this area is Lake Powell, the reservoir that fills the gorge behind Glen Canyon Dam on the Colorado River. Swimming and boating are popular activities, but the scenery, plants, and wildlife also are interesting. Wild flowers, cottonwood, willow, and tamarisk grow along the streams, and waterfowl use the river as a flyway during migrations. The catfish are native, but rainbow trout, kokanee salmon, and largemouth bass have been introduced.

Grand Canyon National Park *(Arizona)*

Few places in the world offer a geological perspective as long and breathtaking as does the Grand Canyon, where the Colorado River has been cutting down through solid rock for several million years. Besides the magnificent Colorado River itself, the reservation includes spectacular waterfalls, springs that ooze from the canyon walls, and several impressive tributary streams flowing through side canyons. The park's unique fish life includes the squawfish, which is the largest minnow in North America, and the knife-shaped humpback sucker. (See pages 186 to 189.)

Grand Teton National Park *(Wyoming)*

Start your exploration of this park with a raft journey down the Snake River. Nesting along its banks are the bald eagle, osprey, Canada goose, screech and horned owls, yellow-bellied sapsucker, red-shafted flicker, brown creeper, and the rare trumpeter swan. In the brush-covered flatlands east of the river the sage grouse struts through its spring courtship rites. Watch for beaver dams along the course of the river and in its tributaries, and be on the alert for a glimpse of the park's famous herd of American elk.

Great Smoky Mountains National Park
(North Carolina, Tennessee)

More than six hundred miles of mountain streams make this a fisherman's paradise. Rainbow trout are plentiful, in addition to brook trout, smallmouth bass, greenside darters, warpaint shiners, dace, chub, sculpins, and about seventy other species. Amphibians also are abundant: the park's watercourses and moist forests harbor many kinds of toads, ten species of frogs, and a variety of salamanders, including thirty-inch-long hellbenders. Watch for spiny softshell turtles in streams at low elevations along the park's northern and western boundaries.

HELLBENDER

Hot Springs National Park *(Arkansas)*

Over a million gallons of water flow each day from the forty-seven hot springs of the park. The origins and workings of the springs, explanations of the 143-degree average temperature of the water, and the geology of the area are depicted in displays at the park's visitor center.

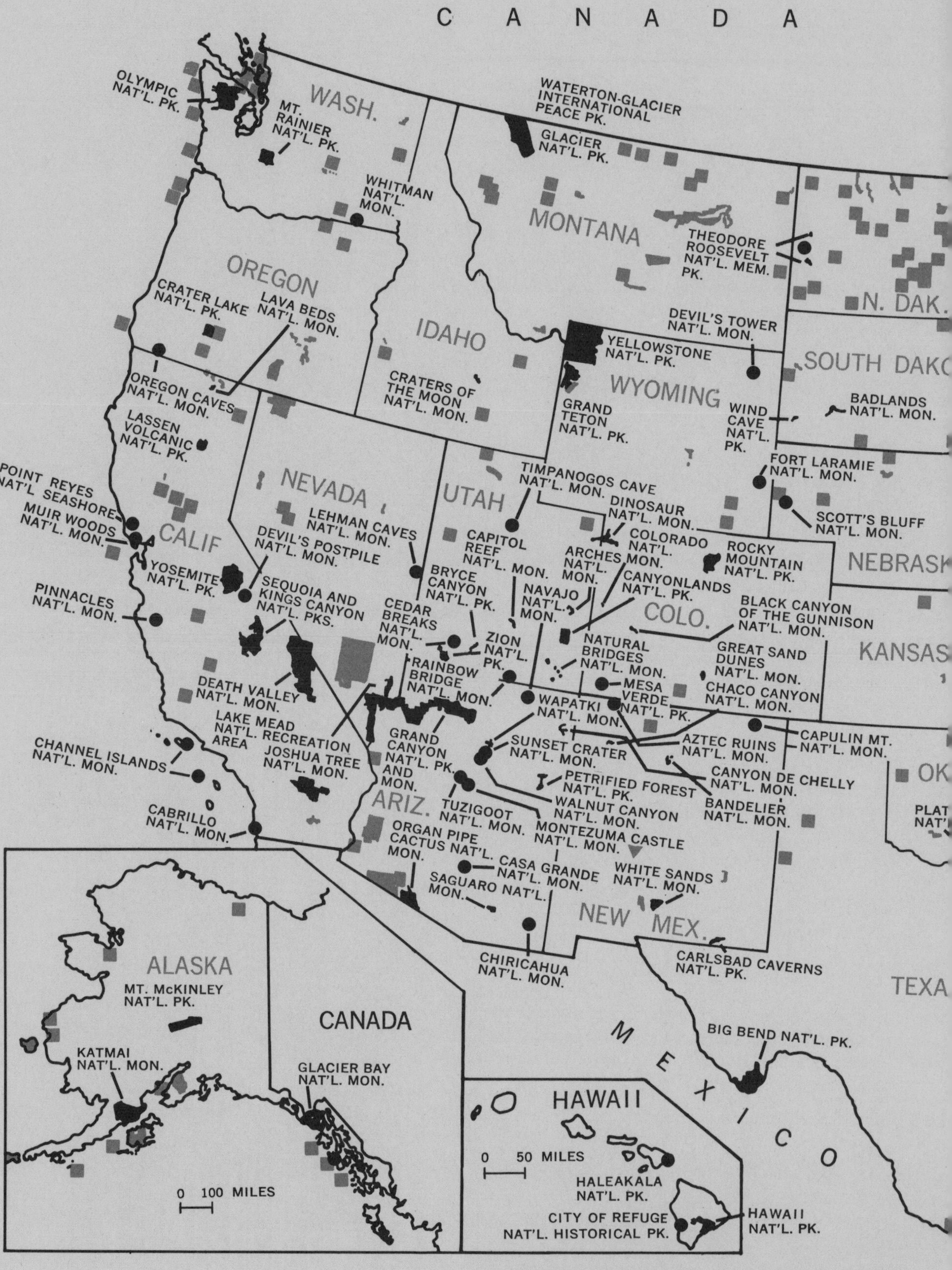

CANADA

WASH.
OLYMPIC NAT'L. PK.
MT. RAINIER NAT'L. PK.
WHITMAN NAT'L. MON.
WATERTON-GLACIER INTERNATIONAL PEACE PK.
GLACIER NAT'L. PK.
MONTANA
THEODORE ROOSEVELT NAT'L. MEM. PK.
N. DAK.
OREGON
CRATER LAKE NAT'L. PK.
LAVA BEDS NAT'L. MON.
IDAHO
DEVIL'S TOWER NAT'L. MON.
YELLOWSTONE NAT'L. PK.
SOUTH DAKO
OREGON CAVES NAT'L. MON.
CRATERS OF THE MOON NAT'L. MON.
WYOMING
GRAND TETON NAT'L. PK.
WIND CAVE NAT'L. PK.
BADLANDS NAT'L. MON.
LASSEN VOLCANIC NAT'L. PK.
POINT REYES NAT'L SEASHORE
MUIR WOODS NAT'L. MON.
NEVADA
UTAH
TIMPANOGOS CAVE NAT'L. MON.
DINOSAUR NAT'L. MON.
FORT LARAMIE NAT'L. MON.
SCOTT'S BLUFF NAT'L. MON.
CALIF.
LEHMAN CAVES NAT'L. MON.
DEVIL'S POSTPILE NAT'L. MON.
CAPITOL REEF NAT'L. MON.
COLORADO NAT'L. MON.
ROCKY MOUNTAIN NAT'L. PK.
NEBRASK
ARCHES NAT'L. MON.
CANYONLANDS NAT'L. PK.
YOSEMITE NAT'L. PK.
SEQUOIA AND KINGS CANYON NAT'L. PKS.
BRYCE CANYON NAT'L. PK.
NAVAJO NAT'L. MON.
COLO.
BLACK CANYON OF THE GUNNISON NAT'L. MON.
PINNACLES NAT'L. MON.
CEDAR BREAKS NAT'L. MON.
ZION NAT'L. PK.
NATURAL BRIDGES NAT'L. MON.
GREAT SAND DUNES NAT'L. MON.
KANSAS
RAINBOW BRIDGE NAT'L. MON.
MESA VERDE NAT'L. PK.
CHACO CANYON NAT'L. MON.
DEATH VALLEY NAT'L. MON.
WAPATKI NAT'L. MON.
LAKE MEAD NAT'L. RECREATION AREA
CAPULIN MT. NAT'L. MON.
GRAND CANYON NAT'L. PK. AND MON.
SUNSET CRATER NAT'L. MON.
AZTEC RUINS NAT'L. MON.
CHANNEL ISLANDS NAT'L. MON.
JOSHUA TREE NAT'L. MON.
OK
PETRIFIED FOREST NAT'L. PK.
CANYON DE CHELLY NAT'L. MON.
ARIZ.
WALNUT CANYON NAT'L. MON.
BANDELIER NAT'L. MON.
PLAT NAT'
CABRILLO NAT'L. MON.
TUZIGOOT NAT'L. MON.
MONTEZUMA CASTLE NAT'L. MON.
ORGAN PIPE CACTUS NAT'L. MON.
CASA GRANDE NAT'L. MON.
WHITE SANDS NAT'L. MON.
SAGUARO NAT'L. MON.
NEW MEX.
CHIRICAHUA NAT'L. MON.
CARLSBAD CAVERNS NAT'L. PK.
TEXA
MEXICO
BIG BEND NAT'L. PK.

ALASKA
MT. McKINLEY NAT'L. PK.
CANADA
KATMAI NAT'L. MON.
GLACIER BAY NAT'L. MON.
0 100 MILES

HAWAII
0 50 MILES
HALEAKALA NAT'L. PK.
CITY OF REFUGE NAT'L. HISTORICAL PK.
HAWAII NAT'L. PK.

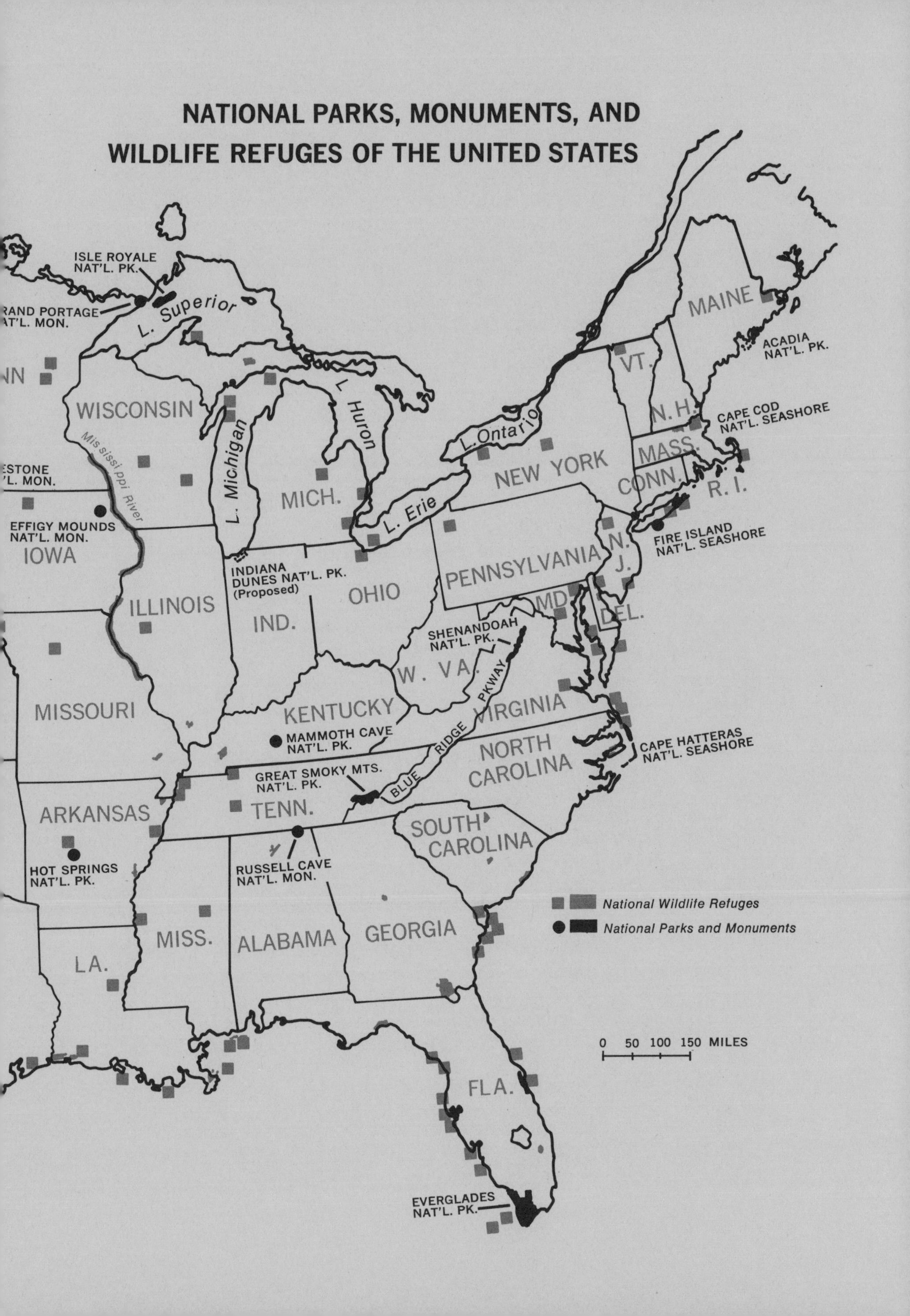
NATIONAL PARKS, MONUMENTS, AND
WILDLIFE REFUGES OF THE UNITED STATES
ISLE ROYALE
NAT'L. PK.
L. Superior
MAINE
ACADIA
NAT'L. PK.
VT.
N. H.
CAPE COD
NAT'L. SEASHORE
MASS.
R. I.
CONN.
NEW YORK
L. Ontario
L. Huron
L. Michigan
WISCONSIN
Mississippi River
MICH.
L. Erie
EFFIGY MOUNDS
NAT'L. MON.
IOWA
FIRE ISLAND
NAT'L. SEASHORE
N. J.
PENNSYLVANIA
INDIANA
DUNES NAT'L. PK.
(Proposed)
OHIO
ILLINOIS
IND.
MD.
DEL.
SHENANDOAH
NAT'L. PK.
W. VA.
MISSOURI
KENTUCKY
VIRGINIA
BLUE RIDGE PKWAY
MAMMOTH CAVE
NAT'L. PK.
CAPE HATTERAS
NAT'L. SEASHORE
NORTH
CAROLINA
GREAT SMOKY MTS.
NAT'L. PK.
ARKANSAS
TENN.
SOUTH
CAROLINA
HOT SPRINGS
NAT'L. PK.
RUSSELL CAVE
NAT'L. MON.
National Wildlife Refuges
National Parks and Monuments
MISS.
ALABAMA
GEORGIA
LA.
0 50 100 150 MILES
FLA.
EVERGLADES
NAT'L. PK.

Katmai National Monument (Alaska)

Among the spectacular sights in this volcanic wilderness are the schools of sockeye and other species of salmon leaping up six- to eight-foot falls as they migrate up the Brooks River to spawn. Salmon fishing is the most popular sport here, both for humans and for the huge Alaska brown bears that gather along streams throughout the summer. Rainbow and Dolly Varden trout, grayling, whitefish, and northern pike also are plentiful.

Lake Mead National Recreation Area (Arizona, Nevada)

Lake Mead, the reservoir behind Hoover Dam on the Colorado River, is one of the largest artificial lakes in the world. Much of the lake is bordered by steep walls of a canyon carved thousands of years ago by the rushing waters of the river. The fish life in this popular recreation area includes the humpback sucker, Colorado squawfish, largemouth bass, rainbow and cutthroat trout, kokanee, channel catfish, black crappie, bluegill, green sunfish, threadfin shad, and mosquito fish. Wild burros, brought here by the early settlers, live along the water's edge together with bobcats, gray and kit foxes, cottontails, and jack rabbits.

Mammoth Cave National Park (Kentucky)

The Echo River flows underground through caves and feeds a surface river, the Green, in this park of caverns and woodland. You can travel along the Echo in a flat-bottomed boat through caves dissolved from limestone by subterranean waters. The creatures of the area, including blindfish, blind salamanders, and other species, are uniquely adapted to life in the underground river. Just outside the park is the sinkhole plain from which water flows into the caves, shaping them and flowing out again into the Green River. The Green flows through hardwood forests filled with a profusion of wild flowers.

BLINDFISH

Mount McKinley National Park (Alaska)

Dominated by Mount McKinley, the highest peak in North America, this park features rugged mountain scenery, several active glaciers, and spectacular wildlife. A number of braided streams—rivers that flow in numerous channels over gravel bars—originate in the mountain glaciers. For the most part, the streams contain too much glacial silt to support fish life, but the arctic grayling is found in some of the clear, cold mountain waters. In addition to moose, wolves, and grizzly bears, the park is a refuge for Barren Ground caribou and heavy-horned Dall's sheep, two species found nowhere else in the National Park System.

Mount Rainier National Park (*Washington*)
Twenty-six active glaciers; beautiful ice caves; broad-floored, steep canyons fashioned by rivers and zone upon zone of wild flowers are a few of the attractions in this snow-capped mountain wilderness. Seasonal fluctuation and glacial silt make most of the streams inhospitable to fish, but elk, deer, and bears are plentiful in the park. Overhead and in the streamside thickets there are chunky slate-gray dippers, woodpeckers, robins, belted kingfishers, warblers, and chickadees.

Olympic National Park (*Washington*)
The mountain peaks of the Olympics afford a panoramic view of the entire water cycle, the movement of moisture from the sea to rivers and streams and back again to the sea. Because of the exceptionally high annual precipitation, several of the park's rivers are bordered by lush temperate rain forests. The most impressive fish here are spawning salmon and five species of trout—cutthroat, rainbow, brook, Dolly Varden, and steelhead. Among the park's mammals are Roosevelt elk, mountain beavers, black-tailed deer, and mountain lions, while the birds include dippers, kingfishers, harlequin ducks, and bald eagles.

Ozark National Scenic Riverways (*Missouri*)
Springs, caves, and forests are scattered along 140 miles of the unspoiled Current and Jacks Fork Rivers in our first national scenic riverway. More than three-fourths of the area is forest, inhabited by white-tailed deer, gray and fox squirrels, raccoons, opossums, and skunks. Ninety-three fish species have been seen in the rivers, including rock, smallmouth, and largemouth bass; walleyed pike; and chain pickerel. Kingfishers, green herons, teal and wood ducks, and wild turkeys live along the waterways. (See pages 146 and 147.)

CHAIN PICKEREL

Rocky Mountain National Park (*Colorado*)
This superb Rocky Mountain wilderness, including 107 named peaks over ten thousand feet high, is dotted with hundreds of icy mountain lakes and scarred by scores of rushing streams. From a vantage point twelve thousand feet up, on Trail Ridge Road, the visitor looks down on alpine meadows in the headwaters basin of the Colorado River. Since the park straddles the Continental Divide, other areas in the park drain eastward into the Atlantic Ocean. Watch especially for beavers and beaver dams, elk, and mountain sheep.

Sequoia and Kings Canyon National Parks (*California*)
In Kings Canyon Park the many-forked Kings River runs between the granite walls of mile-deep chasms, and in Sequoia,

to the south, the Kern and Kaweah Rivers are the major streams. These adjacent parks comprise thirteen hundred square miles of glacier-carved mountain wilderness, where giant sequoias, splendid waterfalls, and two thousand miles of streams are prime attractions. Brook, brown, rainbow, and California golden trout lure anglers to the area. Other wildlife includes the mule deer, the golden eagle, and the black bear, which is dangerous only if you molest it.

GRIZZLY BEAR

Yellowstone National Park *(Wyoming, Montana, Idaho)*
The park is best known as the world's greatest geyser area, with about three thousand geysers, hot springs, and related phenomena, but other attractions are also worth exploring. In addition to scenic beauties such as the falls and canyon of the Yellowstone River, the park is one of the best areas in the country for observing wildlife. Visitors who take advantage of the park's one thousand miles of hiking and riding trails can expect to see black and grizzly bears, mountain sheep, bison, pronghorns, moose, elk, deer, and more than 240 species of birds. (See pages 64 and 65.)

Yosemite National Park *(California)*
This superlative mountain wilderness is one of the best places in the country for observing evidence of the powerful erosive effects of glaciers. The beautiful Yosemite Valley and its bordering sheer granite cliffs were carved by tremendous ice sheets that ground across the area just a few thousand years ago. Enjoy the scenery in the valley, attend a naturalist's program interpreting Yosemite's glacial story, and then hike or drive up into the mountains for a taste of authentic wilderness. (See pages 26 and 27.)

Zion National Park *(Utah)*
The park's best known feature is Zion Canyon, a spectacular gorge carved by the Virgin River. In some places the canyon is so narrow that it is possible to touch both walls at once. Luxuriant thickets of cottonwood, willows, and other trees along the river contrast with the stark beauty of sheer cliffs and multicolored rock formations. Springs at places such as Hanging Gardens and Weeping Rock support lush growths of ferns and water-loving flowers that are rare elsewhere in the arid Southwest. Most of the animals are typical desert species, such as lizards and roadrunners, but don't be surprised if you sight dippers near the river.

How to Study Fresh-water Life

The best way to learn about life in running water is to visit a stream and study the animals in their natural habitat. Any stream shallow enough for wading is a potential research laboratory. With a minimum of equipment, anyone can discover new facts about the habits and interrelations of creatures that live in flowing water.

The best time of year for studying fresh-water life is in summer and early autumn, when the water is clear and the animals are most active. In spring the water usually is too high and too muddy for easy observation, while in winter many of the animals burrow into the bottom and become dormant. Even so, collecting should not stop altogether in the cooler months, for a good many creatures remain active throughout the year. By the same token, observations should be made after dark as well as during the day; some of the most interesting animals, such as certain salamanders, hide beneath rocks by day and emerge to feed only after dark. A waterproof flashlight is helpful for nighttime collecting, but any flashlight held underwater in a tightly closed jar will serve nearly as well.

Many animals can be captured with no equipment at all. Insects that cling to stones or plant stems can simply be picked off by hand. But to catch some of the more active forms, you will need nets and other devices. Since much of the life in streams is small and protectively colored, specimens are more easily examined against a white background, such as a white-enamel kitchen pan or a photo developing tray. Even a pie plate painted white can be useful. A pocket magnifying glass also is indispensable for detailed examination.

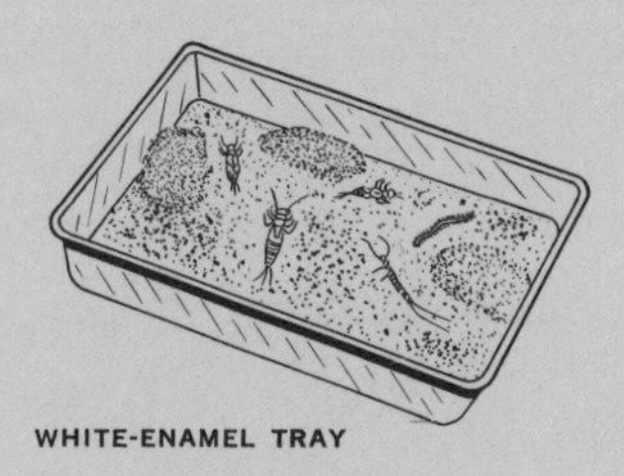

WHITE-ENAMEL TRAY

Although you will want to return most of your specimens to the stream when you have finished observing them, you may want to preserve a few for closer study at home. Small corked vials filled with alcohol are adequate for carrying and preserving them. If you want live specimens, carry them in a plastic jar filled with water. Better still, if you do not have far to go, simply pack the animals in a container filled with moist vegetation.

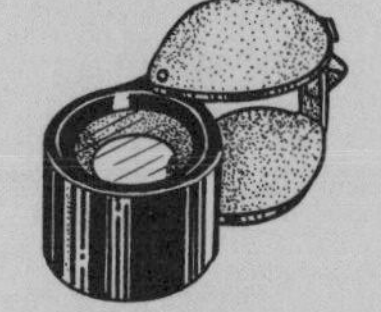

POCKET MAGNIFYING GLASS

Keep a record of when and where you collected your specimens, along with any observations you make about their habits or life histories. Sketches and maps will complete your record. Your notebook, of course, should have a sturdy, waterproof cover.

Some suggestions for making simple, inexpensive collecting and study devices are presented here.

Dip net

To capture insects and other small animals as they swim through the water, use a dip net made from an ordinary wire-mesh kitchen strainer. For best results, tape or wire the strainer to a handle about three feet long; part of a discarded broomstick makes a good handle. If the openings of the mesh are large, line the strainer with cheesecloth.

If you wish, you may construct a larger, more elaborate net by sewing a sturdy wire hoop into the opening of a fine-meshed fabric bag and attaching the net to a wooden handle.

Sweep the net back and forth in the water and empty the contents from time to time into a white tray for closer observation.

Plankton net

The tiny plants and animals of the plankton and the bits of organic material carried downstream with the current are too small to capture in ordinary nets. Instead, take a square of fine silk bolting cloth and sew the two sides together to form a cylinder. Sew one end of the tube to a hoop made from a wire clothes hanger, and tie the other end closed with a string. Three cords tied to the hoop and then attached to a longer cord complete the device.

Hold the net in the stream so that the current sweeps plankton into the opening. (A stone or some other weight on the bottom of the hoop will help keep the opening properly oriented.) After a few minutes, let most of the water drain from the net, untie the bottom, and empty the contents into a shallow dish. Larger animals will be visible through a pocket magnifying glass, but you will find even more life if you examine a drop of the water through a microscope.

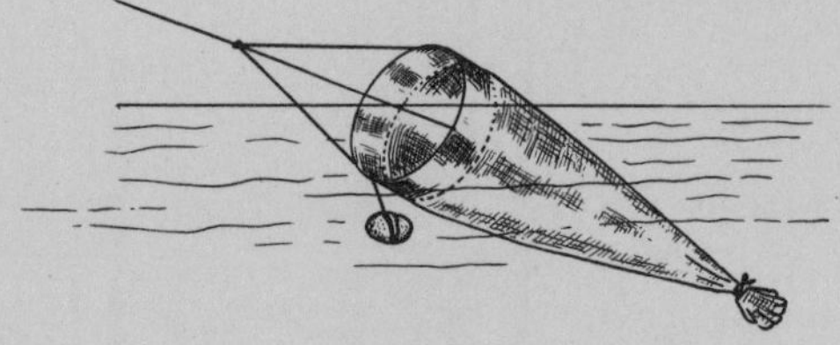

Bottom dredge

To collect samples of plant and animal life from a muddy stream bottom, make a bottom dredge from a cloth bag and a garden rake. Fasten the bag to each end of the rake crosspiece and to a third point a few inches up the rake handle in order to form a more or less triangular opening.

Dragging the rake upstream along the bottom will dislodge a variety of animals, including many forms that burrow into the mud. A dredge of this sort is also effective for collecting the rich assortment of animals that seek the shelter of dense plant growth and would therefore be difficult to reach with an ordinary net. Simply pull the rake across the surface of the plants and let the current wash the animals into the bag.

Waterscope

Underwater life in clear, quiet pools is easily observed from shore, but if the water surface is disturbed by wind or current, you will need some sort of waterscope to eliminate distortion.

To make a simple waterscope, cut a hole in the bottom of a wooden bucket or watertight box and cover the hole with a piece of heavy glass. Tack down strips of wood to hold the glass in place and seal the seams with aquarium cement.

To observe the activities of the underwater animals, plunge the bottom of the waterscope into the water and look through the "window." For best results, work in bright sunlight, but avoid standing where you will cast a shadow on the area you wish to see.

Hand screen

An excellent device for capturing small animals in running water is a hand screen made of wire-mesh window screening tacked to two wooden handles.

Hold the screen across the current, with its bottom edge touching the stream bed, while a companion overturns stones and agitates vegetation on the upstream side. Or you may stand on the upstream side yourself and dislodge stones with your feet. After a few minutes, lift the screen gently from the water, holding it horizontally like a tray, and place it in direct sunlight. Then pick out everything that moves and transfer the creatures to a pan of water for observation. This device is so effective that it may take half an hour to detect all the life in a single collection from a rich stream.

Insectarium

It is difficult to rear stream insects in an aquarium for the purpose of making prolonged studies of their life cycles. Some will survive if you use cool, nonchlorinated water and constantly bubble air through the aquarium. But for better results, you can confine the insects in homemade cages and raise them in their natural habitat.

Make the cage, or insectarium, from a piece of window screening about eighteen inches square. Fold two edges of the screen together to form a cylinder, then flatten one end and fold the edges to seal the bottom.

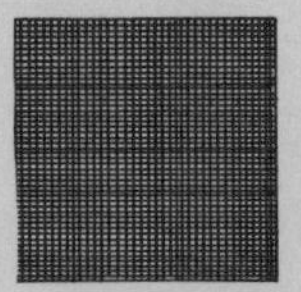

Place one or two live insects in the cage, along with a supply of their natural food, and then close the top. Put the cage in a stream, bracing it in place with stones or sticks. If about one-third of the cage projects above the surface of the water, insects such as dragonfly nymphs will be able to crawl out of the water when they transform into adults. Check the cage regularly and record your observations.

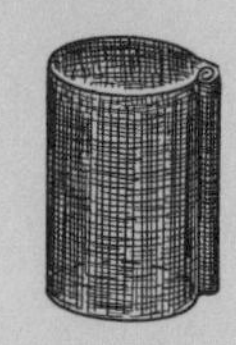

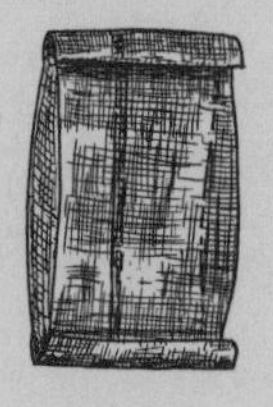

A Guide to Aquatic Insects

Relatively few of the million or so species of insects in the world inhabit rivers and streams. Of those that do, many live in water only during their immature stages; their adult lives are spent on land and in the air. Even so, aquatic insects are usually the most abundant and easily observed forms of life in streams and rivers.

Some, such as water striders, skate across the surface film. Others burrow in muddy bottoms or hide among rocks and fallen logs. Tangled masses of aquatic plants usually teem with insects, and other insects may be observed as they swim through open water. Wherever you look in the stream, in fact, you are likely to find insects involved in their daily struggle to find food and avoid becoming food for other animals.

To help you recognize these fascinating creatures, some of the most prominent forms are illustrated and described here. In the case of insects whose immature and adult stages are conspicuously different, both forms are pictured. Others, such as water boatmen, undergo no great changes in form as they mature, and except for the difference in size the young and adult stages of these insects look practically identical; in such cases, only the adults are illustrated.

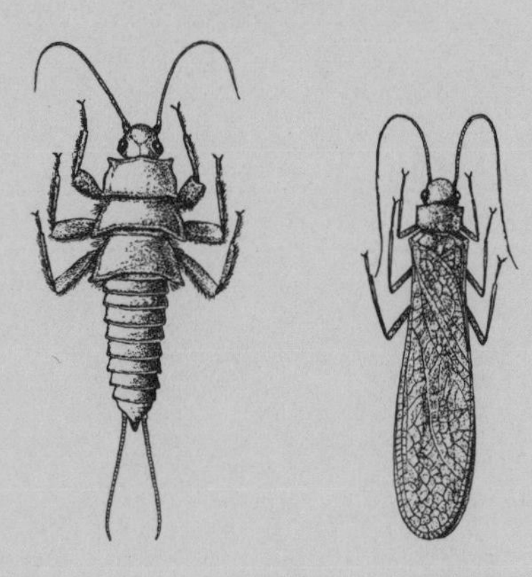

STONEFLY nymphs have two tails, tufts of threadlike gills on their undersides, and two claws on each foot. Most are herbivores. The adults, which look much like nymphs with wings, are usually found clinging to leaves on streamside trees. At rest they fold their wings back over their bodies.

MAYFLY nymphs usually have three tails, gills along their sides, and a single claw on each foot. The fragile-looking, short-lived adults have gauzy, iridescent wings, relatively short antennae, and shrunken mouthparts that resemble receding chins. At rest their long front legs point gracefully forward.

DRAGONFLY nymphs are short, chunky predators with hidden internal gills. The hinged lower jaw folds back under the chin but can be extended forward for grasping prey. The adults spread their four wings horizontally when resting, and the front wings are narrower at the base than are the rear wings.

DAMSELFLY nymphs are more slender than dragonfly nymphs and have three taillike gills, but the mouthparts are similar. The adults usually fold their wings upward and back when resting. The front and rear wings look alike. Unlike dragonflies, whose bulging eyes often meet at the top of the head, damselflies have eyes that are widely separated on the sides of the head.

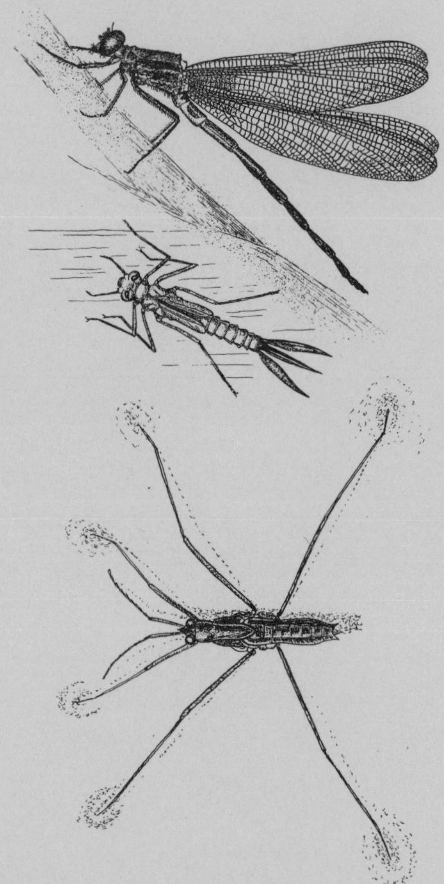

WATER STRIDER nymphs and adults look practically identical. These long-legged predators skate across the surface film and suck juices from prey with their needlelike mouthparts. Some species have wings and can fly away if their pool dries up, but most are wingless.

GIANT WATER BUGS have hidden antennae, paddlelike hind legs, and flattened bodies that are keeled along the underside. They capture prey with their hooked front legs and use their pointed beaks to suck out juices. Males of some species carry incubating eggs on their backs. They often rest at the surface, with the tip of the abdomen projecting from the water for air.

BACKSWIMMERS habitually swim upside down, rowing with their long fringed hind legs. They hang head down from the surface as they take on bubbles of air, then dive and cling to plant stems. These half-inch-long predators are often strikingly patterned. Their boat-shaped bodies are keeled along the back.

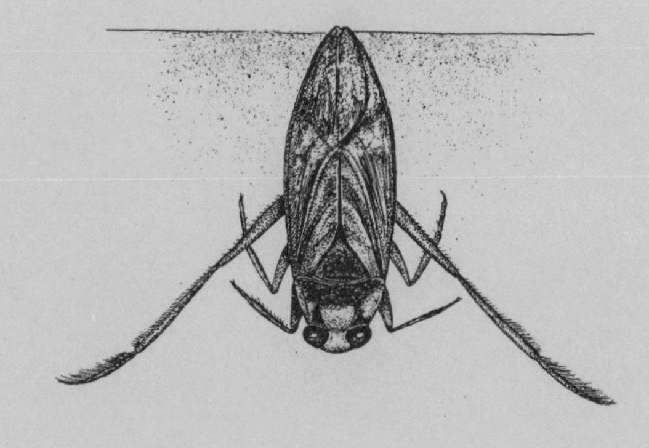

WATER BOATMEN look much like backswimmers, but they swim right side up, moving their fringed hind legs like oars. They use their middle legs for hanging onto rocks and plants and their basketlike front legs for straining microscopic plants from water and mud. Most are dark gray or mottled gray and black.

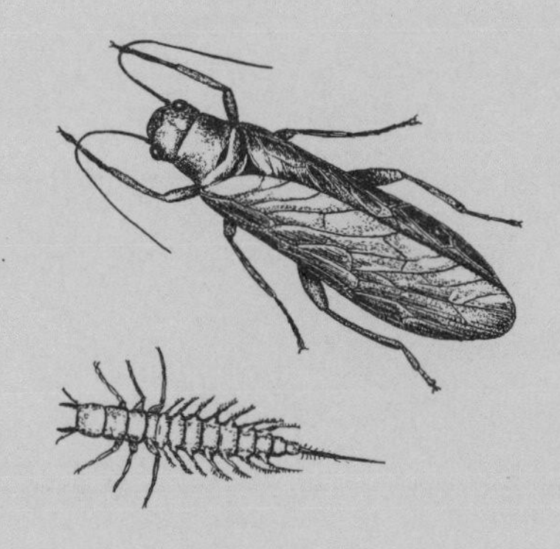

ALDERFLY larvae are brownish bottom-dwelling predators, about an inch long when full-grown. They have long tails and a row of bristly fingerlike gills along each side. The adults are clumsy fliers with membranous wings that lie flat or rooflike over their bodies when at rest. They have long antennae and stout jaws.

DOBSONFLY larvae, or hellgrammites, are two to three inches long when full-grown. They have strong biting mouthparts and a pair of long fingerlike gills on each abdominal segment. The large-winged adults have stout tusklike jaws and long segmented antennae.

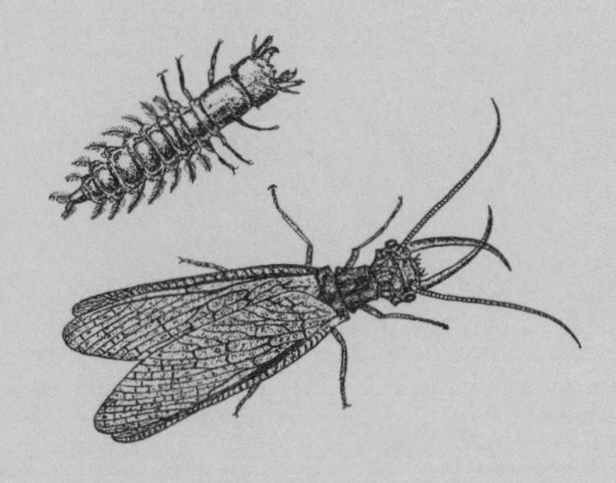

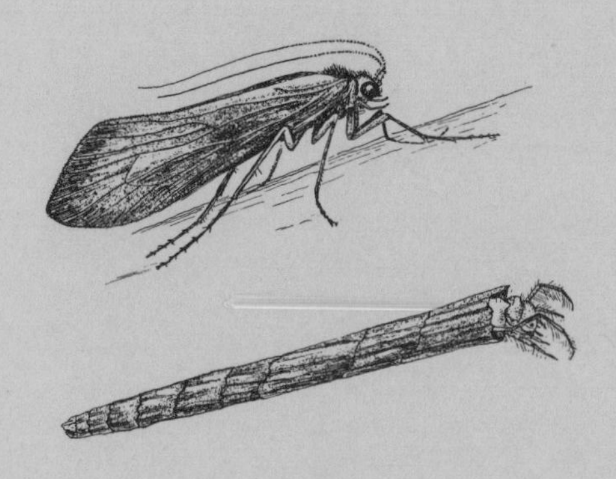

CADDISFLY larvae are inconspicuous omnivores best known for the protective cases of sand, pebbles, or plant materials that most species build. The delicate mothlike adults are usually dull-colored, with long threadlike antennae. The wings are covered with tiny hairs and are typically folded tentlike over the insect's back.

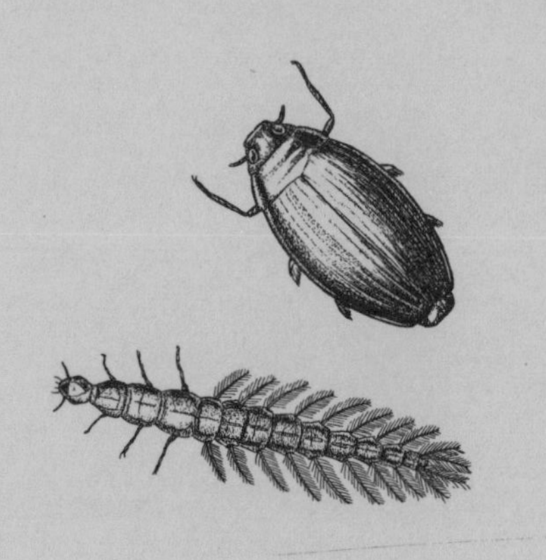

WHIRLIGIG BEETLE larvae are pale, slender predators with hooklike jaws and a row of fringed gills along each side of the body. The flattened, oval adults have eyes divided into upper and lower halves. They gyrate in schools on the surface, paddling with their hind legs. When handled, they give off a fluid that smells like apple seeds.

DIVING BEETLE larvae, or water tigers, grasp prey with their strong, sharp jaws. They often hang from the surface, with the tip of the abdomen exposed to the air. The adults are black or brownish, often with yellow markings. They have slender threadlike antennae and hind legs adapted for swimming. Like the larvae, the adults are predators.

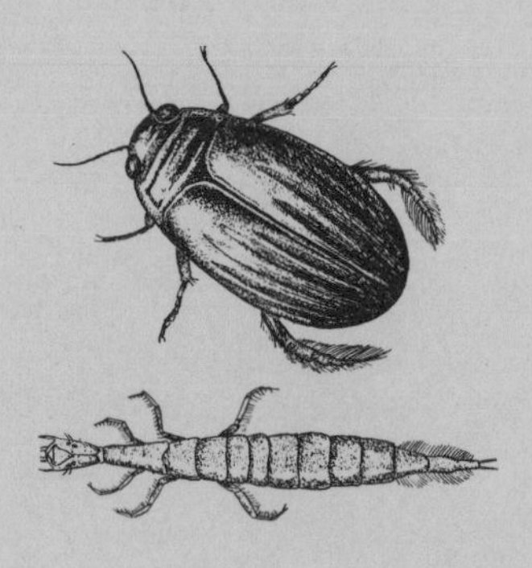

WATER SCAVENGER BEETLE larvae are sluggish predators or herbivores. Most species poke the tip of the abdomen through the water surface for air. The adults resemble diving beetles, but they have short clublike antennae that are often hidden, and they hang tail down instead of head down from the surface when they come up for air.

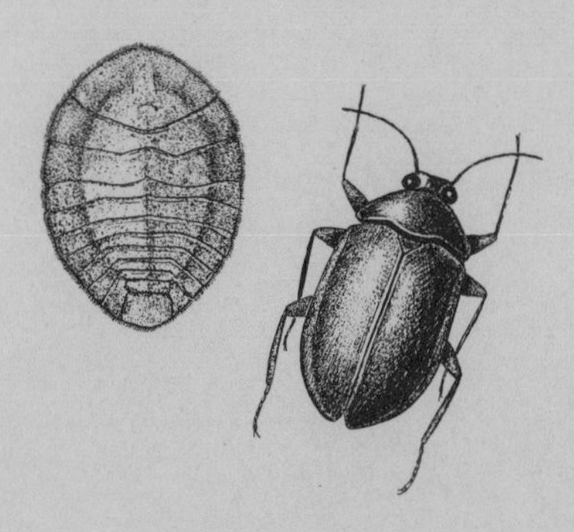

WATER PENNIES are flattened, oval beetle larvae that cling to the undersurfaces of rocks in shallow riffles. The head, legs, and tufts of threadlike gills are visible only when the larvae are pried loose from the stones. The adults are small, inconspicuous beetles covered with fine hairs, and are seen occasionally in flight or creeping over streamside rocks.

MIDGE larvae are slender, fleshy wormlike creatures found in all sorts of aquatic habitats. The adults are delicate flies that are often confused with mosquitoes, but the wings are bare and the females do not bite. Even so, enormous swarms of mating adults are sometimes a nuisance.

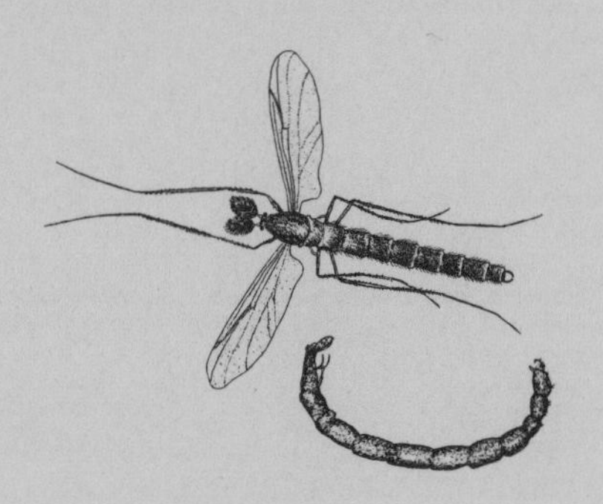

MOSQUITO larvae, also called wrigglers because of their jerky swimming movements, hang head down or parallel to the water surface, with their breathing tubes projecting through the surface film. The adults have characteristic fringes of tiny scalelike hairs on the margins and along the veins of their wings. The females buzz and bite, but the males do not.

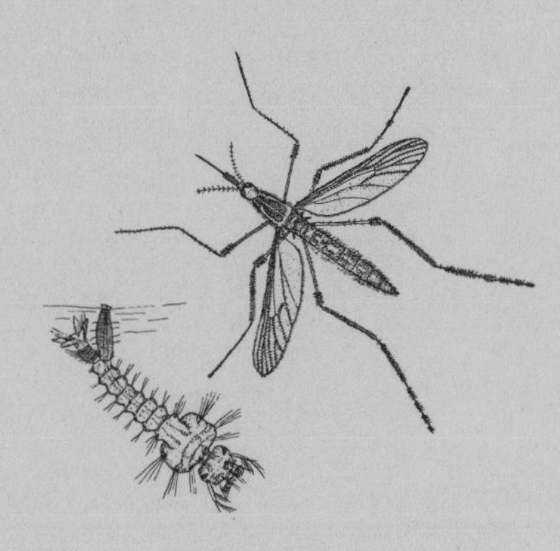

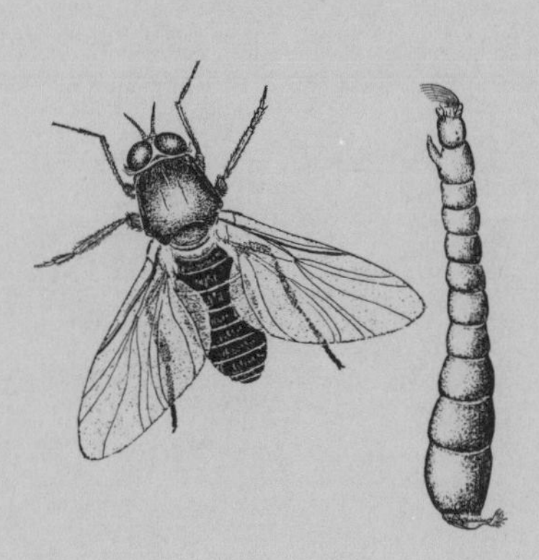

BLACK FLY larvae cling in masses to rocks in swift currents, anchoring their soft bodies with suckers at the hind ends and straining food from the water with fan-shaped mouth brushes. The small black humpbacked adults are notorious pests that inflict painful bites.

Vanishing Fishes

When North America was first colonized, its streams and rivers were filled with fish. The rugged frontiersman had only to drop a line or spread a net, and with scarcely any effort he came up with more fish than he could consume. Today, in far too many of our waterways, the angler may fish for hours and still leave with only a meager catch. Or, if he is successful, he may toss the fish back into the river rather than eat an animal that has spent its life in water tainted with sewage. Instead of trout or shad, he is likely to find that the only fish able to survive in the murky, foul waters of many rivers are carp and other less desirable species.

What is happening to America's fish? In some cases, they have fallen victim to greed. Overfishing has so depleted breeding populations that the fish can barely maintain their numbers. Some of our rivers have been blocked by impassable dams. As a result, migratory species are unable to make their way upstream to their spawning areas. But in most cases, the fish have simply been unable to adapt to the deterioration of their habitat. Because

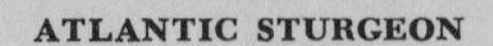

ATLANTIC STURGEON

Like the salmon, the Atlantic sturgeon migrates from the sea to fresh water to spawn. And like the salmon, it fights its way upriver through great stretches of polluted water. But overfishing has been even more important in the decline of this huge fish. Unfortunately, its eggs are valued as caviar, and so many migrating females have been killed for the sake of their eggs that the sturgeon has become extremely rare in coastal rivers from the St. Lawrence to the Gulf of Mexico. Greater protection and better pollution control are needed if we are to save this big, valuable fish.

SHAD

The shad, which occurs from the St. Johns River in Florida to the St. Lawrence in Canada and from southern California to Alaska, is an oceanic fish that spawns in fresh water. Highly valued as food, it has been netted so relentlessly that the annual harvest now is less than one-fifth of what it has been in the past. Pollution and damming of rivers have also taken their toll. Fortunately, under careful management and artificial propagation, this valuable fish now seems to be making a comeback in some of the rivers where it was once plentiful.

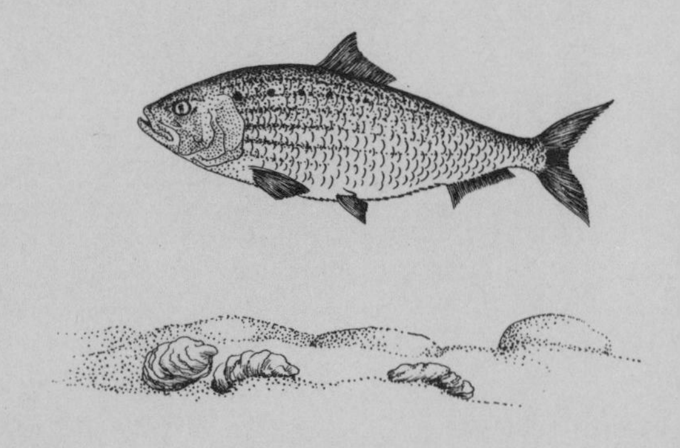

man has mistreated the land, many rivers have become clogged with mud and silt. The water has been fouled with sewage, industrial wastes, poisonous pesticides, and a whole array of other pollutants. Fish—and all other forms of aquatic life—have suffered as a result.

In recent years, several species of fish already have become extinct. The Michigan grayling, the San Gorgonio trout, the thicktail chub, the Pahranagat spinedace—all these are gone forever, and a number of others are poised even now on the brink of extinction. In the case of relict species such as the Olympic mudminnow, extinction would mean a scientific loss; in the case of others, such as the shad, the loss would be a commercial one. But in every case, extinction would be inexcusable. Thoughtful citizens are aware of the plight of these vanishing fishes. We know the reasons for their decline, and we can and must reverse the trend before it is too late. Several species whose existence is in danger are illustrated on these pages.

ATLANTIC SALMON

At one time the Atlantic salmon swam in rivers throughout New England, but today this handsome fish is found only in eight streams in Maine. Dams block many of its migration routes from the sea to the fresh-water streams where it spawns, and many of its gravel spawning beds have been covered with silt. Overfishing also has played a role in the salmon's decline. Some streams are now being stocked with young fish raised in hatcheries. But unless it is given additional help, the Atlantic salmon will remain in grave danger.

OLYMPIC MUDMINNOW

The 2½-inch-long, colorfully patterned Olympic mudminnow is so rare that it was never discovered until 1929. The only western member of the mudminnow family, it has been found only in the Chehalis River drainage basin in western Washington. But as more and more of its swampy spawning areas are drained for farming, it becomes rarer every year. Like a number of other species that occur over very restricted ranges—sometimes in a single spring or stream—the Olympic mudminnow is of great scientific interest as a survivor of an ancient fauna that extended across the continent before the elevation of the Rocky Mountains.

Glossary

Adaptation: An inherited structural, functional, or behavioral characteristic that improves an organism's chances for survival in a particular *habitat*. *See also* Specialization.

Alga (plural *algae*): The simplest of all plant forms, having neither roots, stems, nor leaves. Algae range in size from microscopic single cells to branching forms one hundred feet or more in length.

Annual crop: The total weight or number of living organisms produced in a *community* over the course of a year. The annual crop is used as a measure of the community's *productivity*. *See also* Standing crop.

Antenna (plural *antennae*): A feeler; a sensory appendage that occurs in pairs on the heads of insects and certain other animals.

Aquatic: Living in water.

Bacteria: Simple, colorless one-celled plants, most of which are unable to manufacture their own food using sunlight. Although some bacteria cause diseases, others fill an indispensable ecological role as *decomposers*.

Blood gills: Fleshy protuberances permeated by blood vessels and found on the bodies of many *aquatic* animals. They function as *gills* by allowing gases to pass into and out of the blood as they move across the thin walls of the protuberances.

Canyon: A deep, steep-sided valley eroded into the land by the running water of a stream or river.

Carnivore: An animal that lives by eating the flesh of other animals. *See also* Herbivore.

Carrying capacity: The upper limit on the number of individuals of a particular kind of animal that can be supported within a given unit of *habitat*, such as an area of a stream.

Chlorophyll: A group of pigments that produces the green color of plants; essential to *photosynthesis*.

Community: All the plants and animals in a particular *habitat* that are bound together by *food chains* and other interrelations.

Conservation: The use of natural resources in a way that assures their continuing availability to future generations; the wise use of natural resources.

Consumer: Any living thing that is unable to manufacture food from nonliving substances, but depends instead on the energy stored in other living things. *See also* Decomposers; Food chain; Herbivore; Omnivore; Primary producers.

Continental Divide: The great north-south ridge of the Rocky Mountains in North America; the line of elevated land that separates streams flowing toward the Pacific from those flowing toward the Atlantic.

Creel census: A count of the number and types of fish caught by fishermen in a given area over a specified time period. Information gathered in this way is used by fisheries biologists in determining appropriate stream management methods.

Current: The constant downstream flow of water in a river or stream, resulting from water's tendency to move downslope toward sea level.

Decomposers: Living plants and animals, but chiefly fungi and bacteria, that live by extracting energy from the decaying tissues of dead plants and animals. In the process, they release simple chemical compounds stored in the dead bodies and make them available for use by green plants.

Delta: A deposit of sediments, usually triangular in outline, that collects at the mouths of some rivers.

Diatom: A single-celled *alga* encased in an intricately etched silica shell formed of two halves that fit together like the lid on a box. Diatoms are important *primary producers* in rivers, streams, and other bodies of water.

Drainage pattern: The courses that water follows over a land mass as it flows to the ocean.

Drift: Any or all of the material, living and nonliving, that is carried passively by the current in a river or stream. *See also* Plankton.

Drought: A prolonged period when little or no moisture falls on an area.

Ecology: The scientific study of the relationships of living things to one another and to their *environment.* A scientist who studies these relationships is an ecologist.

Egg: A female reproductive cell. *See also* Fertilization.

Environment: All the external conditions surrounding a living thing.

Erosion: The wearing away of areas of the earth's surface by water, wind, ice, and other natural forces.

Evolution: The process of natural consecutive modification in the inherited makeup of living things; the process by which modern plants and animals have arisen from forms that lived in the past.

Fertilization: The union of a male reproductive cell (*sperm*) with a female reproductive cell (*egg*).

Flash flood: A sudden destructive rush of water across the desert floor after a rainstorm, resulting from the inability of hard-packed desert soil to absorb rain water as quickly as it falls. Besides occurring suddenly, flash floods usually subside quickly.

Flood: A flow of water over an area that is usually dry land.

Flood plain: Low-lying flatlands bordering a river and made up of sediments carried by the stream and deposited during floods.

Food chain: A series of plants and animals linked by their food relationships; the passage of energy and materials from *primary producers* (green plants) through a succession of *consumers.* Green plants, plant-eating insects, and an insect-eating fish would form a simple food chain. *See also* Food web.

Food web: A system of interlocking *food chains.* Since few animals rely on a single food source and since a given food source rarely is consumed exclusively by a single kind of animal, the separate food chains in any natural *community* interlock and form a web.

Fumarole: A vent through which steam and other gases issue from the earth.

Geyser: A hot spring that emits jets of water or steam in intermittent bursts.

Gill: An organ for breathing underwater. *See also* Blood gill; Tracheal gill.

Glacier: A large mass of ice that forms on high ground wherever winter snowfall exceeds summer melting. As snow and ice continue to accumulate at its center, the mass moves slowly downslope until it melts or breaks up.

Habitat: The immediate surroundings (living place) of a plant or animal; everything necessary to life in a particular location except the life itself.

Headwaters: The sources or upper parts of a river.

Hemoglobin: A complex pigment that imparts the red color to blood and functions as a carrier of oxygen in the blood stream.

Herbivore: An animal that eats plants, thus making the energy stored in plants available to *carnivores. See also* Omnivore.

Host: A living organism whose body provides food or living space for another organism. *See also* Parasite.

Incomplete metamorphosis: The type of life history, characteristic of certain insects such as dragonflies and true bugs, in which there is no inactive *pupal* stage. Instead, the immature insect, or *nymph*, undergoes a series of *molts* as it grows, and emerges as a mature adult on its final molt. *See also* Larva; Metamorphosis.

Larva (plural *larvae*): An active immature stage in an animal's life history, during which its form differs from that of the adult. The caterpillar, for example, is the larval stage in the life history of a butterfly; the tadpole is the larva of a frog. *See also* Incomplete metamorphosis; Metamorphosis; Nymph; Pupa.

Lung: A breathing organ consisting of air sacs lined by moist membranes that are permeated by minute blood vessels. Oxygen from the air passes through the membranes into the blood, while carbon dioxide passes from the blood into the air sacs. *See also* Gill.

Metamorphosis: A change in the form of a living thing as it matures, especially the transformation from a *larva* to an adult. *See also* Incomplete metamorphosis; Nymph; Pupa.

Microorganism: A plant or animal too small to be seen without magnification.

Molt: To shed or cast off a body covering such as the shell-like external skeleton of an insect.

Nymph: The immature, preadult form of an insect, such as a dragonfly, whose life history is characterized by *incomplete metamorphosis.* Nymphs, after hatching, live and feed in water until they *molt* and become adults. *See also* Larva; Metamorphosis.

Omnivore: An animal that habitually eats both plants and animals. *See also* Carnivore; Herbivore.

Organic: Pertaining to anything that is or ever was alive or produced by a living plant or animal.

Parasite: A plant or animal that lives in or on another living thing (its *host*) and obtains part or all of its food from the host's body.

Photosynthesis: The process by which green plants convert carbon dioxide and water into simple sugars. *Chlorophyll* and sunlight are essential to the series of complex chemical reactions involved in the process.

Plankton: The minute plants and animals that float or swim near the surface of a body of water. Plant plankton and animal plankton provide an important food source for many *aquatic* animals. *See also* Drift.

Pollution: The fouling of water or air with sewage, heat, industrial wastes, or other contaminants, making them unfit to support many forms of life.

Predator: An animal that lives by capturing other animals for food.

Primary producers: Green plants, the basic link in any *food chain.* By means of *photosynthesis,* green plants manufacture the food on which all other living things ultimately depend. *See also* Consumer.

Production pyramid: A representation of the diminishing amount of *organic* material produced at each successive level along a *food chain.* The decline in *productivity* results from the constant loss of energy through living processes along the food chain. *See also* Pyramid of numbers.

Productivity: The total weight or number of living things produced in a functioning natural *community.* Productivity depends on the interaction of the life and the *environment.*

Pupa (plural *pupae*)**:** The inactive stage in the life of certain insects, during which the *larva* undergoes a gradual reorganization of its tissues in the process of transforming into an adult. *See also* Metamorphosis.

Pyramid of numbers: A representation of the normally declining number of individuals at each successive level on a *food chain. See also* Production pyramid.

Riffle: A rapid; an area of a stream where shallow water races downslope over a bed of stones and gravel.

Scavenger: An animal that eats the dead remains and wastes of other animals and plants.

Snowpack: An accumulated mass of snow, as on a mountainside. In effect, a snowpack is a natural storage place of water, since its gradual melting often feeds streams well into summer.

Spawn: To shed reproductive cells; refers to animals such as salmon and other fishes that shed *eggs* and *sperm* directly into the water.

Specialization: The sum of the *adaptations* that enable a plant or animal to survive in a particular *habitat* or equip it for a particular mode of life.

Species (singular or plural)**:** A group of plants or animals whose members breed naturally only with each other and resemble each other more closely than they resemble members of any similar group.

Sperm: A male reproductive cell. *See also* Fertilization.

Spiracle: An opening for breathing, such as the external openings to an insect's *tracheal* system or the opening through which a tadpole expels water as it breathes.

Spring: A flow of underground water issuing naturally out of the earth.

Standing crop: The total weight or number of living organisms (or of a particular kind of organism) present in a given area at any one time. *See also* Annual crop.

Streamlined: Having a form or body shape that offers a minimum of resistance to air or water.

Surface tension: A property of liquids that causes the surface of a standing body of a liquid to act like an elastic film. Surface tension results because molecules of the liquid at the surface have a stronger attraction for each other than they do for the air above.

Sustained yield: The continuing yield of a forestry or fishery crop achieved by controlling the harvest in a way that assures a steady optimum yield.

Trachea (plural *tracheae*)**:** A minute branching air tube that distributes air through an insect's body. Tracheae are the basic elements in the breathing system of insects and certain of their relatives.

Tracheal gills: Thin-walled fingerlike or platelike *gills* permeated by *tracheae* and found on the bodies of certain *aquatic* insects. Gases pass between the water and the air tubes by moving across the thin membranes that form the walls of the gills.

Transpiration: The process by which water evaporates from plant tissues.

Tributary: A stream that flows into a larger stream or some other body of water.

Watershed: The area drained by a river or stream and all its *tributaries.*

Water table: The upper level of the underground reservoir of water, below which the soil and all cracks and channels in the rocks are saturated. The water table in different areas may lie at the surface or hundreds of feet underground and, depending on rainfall and the rate of removal of the water, may fluctuate from time to time in any given area.

White water: Swiftly flowing, frothy water in a rapid.

Bibliography

FRESH-WATER BIOLOGY

BROWN, E. S. *Life in Fresh Water.* Oxford University Press, 1955.

COKER, ROBERT E. *Streams, Lakes, Ponds.* University of North Carolina Press, 1954.

EDMONDSON, W. T. (Editor). *Fresh-water Biology.* Wiley, 1965.

FREY, DAVID G. (Editor). *Limnology in North America.* University of Wisconsin Press, 1963.

LAGLER, KARL F. *Freshwater Fishery Biology.* William C. Brown, 1956.

MACAN, THOMAS T. *Freshwater Ecology.* Wiley, 1963.

MACAN, THOMAS T., and E. B. WORTHINGTON. *Life in Lakes and Rivers.* Collins, 1951.

NEEDHAM, JAMES G., and PAUL R. NEEDHAM. *A Guide to the Study of Fresh-water Biology.* Holden-Day, 1962.

POPHAM, EDWARD J. *Some Aspects of Life in Fresh Water.* Harvard University Press, 1961.

REID, GEORGE K. *Ecology of Inland Waters and Estuaries.* Reinhold, 1961.

RUTTNER, FRANZ. *Fundamentals of Limnology.* University of Toronto Press, 1963.

WELCH, PAUL S. *Limnology.* McGraw-Hill, 1952.

ANIMAL LIFE

JONES, J. W. *The Salmon.* Harper & Row, 1960.

KLOTS, ELSIE B. *The New Field Book of Freshwater Life.* Putnam, 1966.

LAGLER, KARL F. JOHN BARDACH, and ROBERT R. MILLER. *Ichthyology.* Wiley, 1962.

MORGAN, ANN HAVEN. *Field Book of Ponds and Streams.* Putnam, 1930.

MURIE, OLAUS J. *A Field Guide to Animal Tracks.* Houghton Mifflin, 1954.

NEEDHAM, JAMES G., J. R. TRAVER, and YIN-CHI-HSU. *The Biology of the Mayflies.* Comstock, 1935.

NEEDHAM, JAMES G., and MINTER WESTFALL. *Manual of the Dragonflies of North America.* University of California Press, 1955.

OLIVER, JAMES A. *The Natural History of North American Amphibians and Reptiles.* Van Nostrand, 1955.

PALMER, RALPH S. *The Mammal Guide.* Doubleday, 1954.

PENNAK, ROBERT W. *Fresh-water Invertebrates of the United States.* Ronald Press, 1953.

POUGH, RICHARD H. *Audubon Land Bird Guide.* Doubleday, 1949.

POUGH, RICHARD H. *Audubon Water Bird Guide.* Doubleday, 1951.

SCHRENKEISEN, R. *Field Book of Fresh-water Fishes of North America.* Putnam, 1963.

SCHULTZ, LEONARD P., and EDITH M. STERN. *The Ways of Fishes.* Van Nostrand, 1948.

USINGER, ROBERT L. (Editor). *Aquatic Insects of California.* University of California Press, 1956.

WATER PLANTS

FASSETT, NORMAN C. *Manual of Aquatic Plants.* University of Wisconsin Press, 1957.

PRESCOTT, G. W. *How to Know the Fresh Water Algae.* William C. Brown, 1964.

SMITH, GILBERT M. (Editor). *The Fresh-water Algae of the United States.* McGraw-Hill, 1950.

ECOLOGY

BENTON, ALLEN H., and WILLIAM E. WERNER, JR. *Field Biology and Ecology.* McGraw-Hill, 1966.

BUCHSBAUM, RALPH, and MILDRED BUCHSBAUM. *Basic Ecology.* Boxwood Press, 1957.

ODUM, EUGENE P., and HOWARD T. ODUM. *Fundamentals of Ecology.* Saunders, 1959.

SMITH, ROBERT LEO. *Ecology and Field Biology.* Harper & Row, 1966.

CONSERVATION

CARR, DONALD E. *Death of the Sweet Waters.* Norton, 1966.

CLAPP, GORDON R. *The T.V.A.: An Approach to the Development of a Region.* University of Chicago Press, 1955.

HELFMAN, ELIZABETH F. *Rivers and Watersheds in America's Future.* McKay, 1965.

HYNES, H. B. N. *The Biology of Polluted Waters.* Liverpool University Press, 1963.

KLEIN, LOUIS. *Aspects of River Pollution.* Academic Press, 1957.

U.S. DEPARTMENT OF HEALTH, EDUCATION AND WELFARE. *Clean Water: A Challenge to the Nation.* Public Health Service Publication No. 816, 1960.

WATER RESOURCES POLICY COMMISSION. *Ten Rivers in America's Future: The Report of the President's Water Resources Policy Commission,* Vol. 2. U.S. Government Printing Office, 1950.

GENERAL READING

AMOS, WILLIAM HOPKINS. *Life in Flowing Waters.* Doubleday, 1960.

BARDACH, JOHN. *Downstream: A Natural History of the River.* Harper & Row, 1964.

BUTCHER, DEVEREUX. *Exploring Our National Parks and Monuments.* Houghton Mifflin, 1960.

DAVIS, KENNETH S., and JOHN ARTHUR DAY. *Water: The Mirror of Science.* Doubleday, 1961.

EIFERT, VIRGINIA S. *River World: Wildlife of the Mississippi.* Dodd, Mead, 1959.

LEYDET, FRANÇOIS. *Time and the River Flowing.* Sierra Club, 1964.

MILLER, M. R. *The Brook Book.* Doubleday, 1904.

PERRY, JOHN, and JANE GREVERUS PERRY. *Exploring the River.* McGraw-Hill, 1960.

POWELL, JOHN WESLEY. *Exploration of the Colorado River.* University of Chicago Press, 1957.

THOREAU, HENRY D. *The Maine Woods.* College and University Press, 1965.

U.S. DEPARTMENT OF AGRICULTURE. *Water: The Yearbook of Agriculture, 1953.* U.S. Government Printing Office, 1953.

Illustration Credits and Acknowledgments

COVER: Alaska brown bear with salmon, Leonard Lee Rue

ENDPAPERS: Peter G. Sanchez

UNCAPTIONED PHOTOGRAPHS: *8–9*: Forest and stream, J. E. Coufal *58–59*: Mountain stream, James W. Larson *108–109*: Snowy egret, Bill Ratcliffe *150–151*: Aerial view of Potomac River, Washington, D.C., Frank McGuire, courtesy of *National Wildlife* magazine

ALL OTHER ILLUSTRATIONS: *10–11*: G. Ronald Austing from Franklin W. Lane, Ltd. *12*: Patricia C. Henrichs *13*: J. M. Conrader *14*: B. B. Jones *15*: Patricia C. Henrichs *16–17*: Edward S. Ross *18–19*: Lloyd Tevis, Jr. *20*: Mark Binn *21*: Larry West from Full Moon Studio *22*: Edward S. Ross *23*: John W. Evans *25*: Grant Heilman *26–27*: James W. Larson *28*: Robert L. Usinger *29*: Allan Roberts *30*: B. B. Jones *31*: Roland Wauer *32–33*: Grossman from Freelance Photographers Guild *34*: Leonard Lee Rue from Freelance Photographers Guild *35*: Jane Burton from Photo Researchers *36*: B. B. Jones *37*: Edward S. Ross *38*: Patricia C. Henrichs *39*: Treat Davidson from National Audubon Society *40*: H. Eisenbeiss from Photo Researchers *41*: Marjorie Pickens *42*: Allan Roberts *43*: Edward S. Ross *44–45*: Charles Fracé *46–47*: Patricia C. Henrichs *48–51*: Edward S. Ross *52–53*: Robert Strindberg *54*: Edward S. Ross *55*: Larry West from Full Moon Studio *56*: Robert Strindberg *60*: David Mohrhardt from Full Moon Studio *61*: Stephen Collins *62*: M. Woodbridge Williams *63*: Grinnell from Monkmeyer Press Photos *64*: Leonard Lee Rue; W. J. Schoonmaker *65*: Ansel Adams *66–67*: Robert Clemenz *68–69*: Glenn D. Chambers *70–71*: Hans Zillessen from G.A.I. *72*: Josef Muench *73*: R. S. Simmons *74*: Charlie Ott *75*: Robert C. Hermes *76*: Harry Engels *77*: Michael Wotton *78*: Bob and Ira Spring *79*: Peter G. Sanchez *80*: NASA *82–83*: Ernest Gay *84*: Charles Fracé *85*: Simons from Photo Researchers *86–87*: Elgin Ciampi *88–89*: Leonard Lee Rue *90–91*: Mark Binn *93*: Allan Roberts *94–95*: Edward S. Ross *96*: John Gerard *97*: Stanley Jewett, Jr. *98*: Grant Haist *99*: John and Jane Perry *100*: Norman Lightfoot; Allan Roberts *101*: John W. Evans *102*: Charles Fracé *103*: Freelance Photographers Guild *104*: Treat Davidson from National Audubon Society *105*: Charles Fracé *106*: NASA *110*: A. Assid from Full Moon Studio *111*: Robert W. Mitchell; B. B. Jones *112*: Robert W. Mitchell *113*: B. B. Jones *114*: Bill Ratcliffe *115*: Patricia C. Henrichs *116–117*: Charles Fracé *118*: Lawrence Pringle *119*: Patricia C. Henrichs *120*: Mark Binn *121*: Charles Fracé *122–123*: Elgin Ciampi *124–125*: Hans Zillessen from G.A.I. *126*: Ray Atkeson *128*: Edward S. Ross *129*: Paul R. Needham *131*: Mark Binn *132*: Grant Heilman *134*: Les Line *135*: Grant Heilman *136–137*: Bill Strode *138–139*: Charles Fracé *140*: Mark Binn *141*: Dick Kent Photography *142–145*: Bill Strode *146–147*: Don Wooldridge *148*: Ernest Gay *152–153*: Foldes from Monkmeyer Press Photos *154*: Lawrence Pringle *155*: Mark Binn *156–157*: Graphic Arts International *158–159*: Freelance Photographers Guild *160*: Vernon C. Applegate, Bureau of Commercial Fisheries *161*: Hans Zillessen from G.A.I. *162–165*: John Gerard *166*: Don Wooldridge *167*: United Press International *168*: Thase Daniel *169*: Matthew Vinciguerra; Jack Dermid *170–171*: Dick Kent Photography *173*: John Crawford *174–175*: Grant Heilman *176*: Mark Binn *177*: N. A. Bishop *178–179*: Cletis Reaves from Alpha Photo Associates *180–182*: Josef Muench *183*: Graphic Arts International *185*: Ruth Kirk *188–189*: Philip Hyde *190*: Steven C. Wilson from D.P.I. *191*: Evan J. Davis *192*: Charles Fracé *193*: Leonard Lee Rue *194*: John Crawford *194–195*: John Crawford; Leonard Lee Rue *196*: John Crawford *197*: Mark Binn *198*: Bill Strode *203–205*: Charles Fracé *206–207*: Graphic Arts International *208–210*: Charles Fracé *211–214*: Mark Binn *215–219*: Charles Fracé *220–221*: Patricia C. Henrichs (Olympic mudminnow after Dorothea B. Schultz)

PHOTO EDITOR: ROBERT J. WOODWARD

ACKNOWLEDGMENTS: *The author gratefully acknowledges the generosity of Donald Seegrist, who permitted him to use data contained in an unpublished doctoral thesis. The publisher wishes to thank O. L. Wallis, William Perry, and Douglass Hubbard of the National Park Service, all of whom read the entire manuscript and offered valuable suggestions. Dr. Leonard P. Schultz of the Smithsonian Institution and Gerald E. Holsinger, Assistant Librarian at the American Museum of Natural History, also contributed valuable assistance in locating reference material for illustrations.*

Index

[Page numbers in **boldface** type indicate reference to illustrations and maps.]

Adaptations, 222
Adirondack Mountains, 155
Air bladders, 112
Alaska brown bears, **192, 197**
Alderflies, **217**
Alders, 119
Algae, 13, 22, **40,** 59, 79, 113–117, **121,** 123, 127, 222
  in hot springs, 63, **66–67**
  in polluted waters, **134,** 139
  as primary producers, 110, 115–**116,** 123–125, 127
Allegheny River, 165
Alligator snapping turtles, 102
Amphibians, 40, 47 (*see also* Frogs; Salamanders)
Animals, 22, 69
  cave, 73–74
  changing habitats of, 18–19
  microscopic, 13, 90, 115, 117, 192
  tracks of, 30–31, **84**
  (*See also specific animals*)
Annual crop, 121, 124, 222
Antennae, 222
Aphids, 119
Appalachian Mountains, 158
Aquatic, defined, 222
Arctic Circle breeding grounds, 75, 151
Arkansas River, 163, 166
Army Corps of Engineers, 166
Arrowheads, 112
Atlantic Ocean, 158, 176
Atlantic salmon, **221**
Atlantic sturgeon, **220**
Atomic Energy Commission, 143
Azolla (water ferns), 36

Backswimmers, **217**
Bacteria, 13, 115–116, 119, 222
  purification by, 138–140
Bass, 103, 105, 138, 140, 169, 172
  in food chains, **116,** 123–125
  as gill breathers, **39**
Bats, 98, 116
Bear Mountain State Park (New York), **152–153**
Bears, 64, 155, 191, **193–197, 210**
  tracks of, 30–31
Beavers, **34**–35, 39
  lodges of, 31, 34
Big Bend National Park (Texas), 172, **174–175,** 203–204
Big Spring State Park (Missouri), **72**
Bighorns, 64, **177,** 183
Birds, 21, 86, 192 (*see also specific birds*)
Bison, 64
Bitterns, **168**
Black bears, 31, 155
Black-fly larvae, **15–17,** 92, 117, **121, 219**
Blindfish, **208**
Blobs (sculpins), 20, 39, **46,** 155
Blood gills, 44, 222
Bloodworms, 44, 97
Blue crabs, **105**
Bobcats, 172
Bonneville Dam, 190
Bowfins, **122–123**
Breathing, underwater, 38–47
  gill, 39, 42, 44–47, 135
  in polluted water, 135–136, 138–139
Brook trout, 129
Brown bears, **192–195, 197**
Brown trout, 129, 130
Bubbles, air, 42–44
Buffalo fish, 105
Bullheads, 103–105, 124–**125,** 172
Bur reeds, 112
Burrowers, 92, **161, 185**

Cactuses, 172, 176, 183
Caddisflies, 119, **218**
  larvae of, **12–13, 24,** 28, 45, 48, 57, 66, 92, 117, 138, 140, **218**
  metamorphosis of, **48–51**
California, University of (Berkeley), 128–131
Canyons, 183–184, 222
Carbon dioxide, 38–39, 110
Cardinal flowers, **61**
Carnivores, 116, 125, 222
Carp, 102–103
Carrying capacity, 131, 222
Cascades, **41**
Catbirds, 59
Catfish, 103–105, 124–**125,** 172
Cathedral in the Desert (Arizona, Utah), **182**
Cattails, 89, 95, 109, **111**–112, 168
Caves, 73–74, **79**
Caviar, 190, 220
Celilo Falls, **191**
Chain pickerel, **209**
Channel catfish, **104**
Chinook salmon, 191
Chisos Mountains, 172
Chlorophyll, 110, 222
Chum salmon, 191–192
Cladophora (algae), **115, 121**
Clams, 107
Clearwater River, 149
Clouds, formation of, 68–71
Coho salmon, 191–192
Colorado River, **106,** 177–184
Columbia River, **126,** 177, 184–185, 190–199
  salmon in, 105, 190–192, 197, 199
Columbine, 72
Community, defined, 222
Conservation, 135, 143, 169, 222
Consumer, 116, 124, 222
Continental Divide, *map* **176,** 177, 222
Cormorants, 107
Cottonwoods, 83–84, 172, **203**
Coulee Dam National Recreation Area (Washington), 204
Coyotes, 172
Crabs, **105,** 107
Crappies, 105, 169, 172
Crayfish, 20–**21, 46,** 84, 99, 116
  cave, 73–74
Creel censuses, 120, 130, 222
Crevasses, 78
Croakers, 107
Current, 46, 223
Current River, 145

Dace, 155
Daddylonglegs, 79
Dalles Dam, 191
Dams, 92, 145, 183–184, 190–192, 197
  flood-control, 166–167, 169, **180**
Damselflies, **94–95, 216**
  nymphs of, 22–**23,** 45, **216**
Darters, 20
Davis Dam, 180, 183
Death Valley (California), 62, 67
Decay, 109–110, 116
Decomposers, 116, 119, 127, 223
Deer, **31,** 122, 155, 172
Deer mice, 61
Delaware River, 149, 158
Deltas, 166, 176, 223
Deserts, 100, 171–172, 175
  along Colorado River, 177–179, 182–184
  (*See also* Death Valley)
Detergents, 136, 141–142
Devils River, 176
Diatoms, 115, 117, 123, 223
Dinosaur National Monument (Colorado, Utah), 204
Dip nets, **212**
Dippers, **28**–29
Diseases, 140, 169
Diving beetles, **218**
Dobsonfly larvae (hellgrammites), **22, 121, 217**
Donacia beetles, 43–**44**
Dragonflies, 12, 98, **216**
  mating of, 94–95
  metamorphosis of, 49–50, **52–53**
  nymphs of, 22–23, 92, 94, **216**
Drainage basins (*see* River basins)
Drainage pattern, 223
Dredges, bottom, **213**
Drift, 117, 223
Droughts, 19, 180, 223
Ducks, 75, 102, 120, 183
  migration of, 169–170
Duckweeds, 36, **111,** 113
Dytiscus (diving beetles), **44**

Eagle Creek "Punch Bowl" (Oregon), **126**
Eagles, 143, 191
Earthworms, 92, 139
Ecology, 223
Eelgrass, **122–125**
Eels, 105
Egg, defined, 223
Egrets, **108,** 183
Electric power, 143, 167, 169, **180–181**
Elephant Butte Reservoir, 172
Elks, **64**

Energy, 110, 115, 127
Environment, defined, 223
Erosion, 83, 131, 135, 169, 172, 223
  and formation of canyons, 183–184
Estuaries, **106,** 155
Evaporation, 68–71
Evolution, 223
Extinction, threatened, 75, 172, 190
Eyes, **38**
  animals without, 73–74

Feldspar Brook (New York), **154**
Ferns, 36, 59, 72, 185
Fertilization, 223
Fish, 12, 23, 98, 102–105, 109
  caught by birds, 9–11, 98–99
  cave, 73–74, **208**
  creel censuses of, 120, 130
  in hot springs, 66–67
  in otters' diets, 84, 88
  parasites on, **91–92, 160–161**
  pollutants and, 135–136, 139, 142–143, 197
  prey of, **21,** 51, 97
  vanishing, **220–221**
  (*See also specific fish*)
Fish hatcheries, 128
Fish ladders, 192, **197**
Flash floods, **178–179,** 223
Flatworms, 61, 73
Fleas, glacier, 79
Flood plain, 223
Floods, 76, **166–167,** 169, 180, 223
  effect of, on habitats, 18–19
  flash, **178–179,** 223
Flounders, 107
Fog, **68–69,** 151
Fontinalis (fountain moss), 113
Food chains, 115–119, 143, 223
  measurement of, 121–125, 127
Food webs, 116, 119, 223
Forests, 74–75, 84, 100, 172, 190
Fountain moss, 113
Foxes, 175
Fresh-water life, how to study, 211–214
Frogs, 84, 98–99, 102, 120 (*see also* Tadpoles)
Fumaroles, 62, 223
Fungi, 119, 138

Gar, 105, 123–125, 127
Geese, 75, 151, 170, 183
Geysers, 62–63, 66, 223
Gills, 39, 42, 90–91, 135, 223
  blood, 44, 222
  insect, 44–47, 92
  tracheal, 226
Glacier fleas, 79
Glacier National Park (Montana), 76, 204
Glaciers, 69–71, 75–79, 223–224
Glen Canyon Dam, 183
Glen Canyon National Recreation Area (Arizona, Utah), 204–205
Glossosoma (caddisflies), **13**
Goldfish, 39
Grackles, 61
Grand Canyon National Park (Arizona), **148,** 184, **186–189,** 203, 205
Grand Canyon of the Yellowstone, 65
Grand Coulee Dam, 190
Grand Falls, **178–179**
Grand Teton National Park (Wyoming), 205
Grasshoppers, 38–39
Great Basin, 176
Great blue herons, **98–99**
Great Lakes, 159–161
Great Smoky Mountains, 30
Great Smoky Mountains National Park (North Carolina, Tennessee), 205
Grebes, 143
Green River, 145
Gulf of California, **106,** 176, 184
Gulf Coast, 75, 102
Gulf of Mexico, 163–166, 170, 176
  (*see also* Gulf Coast)
Gulls, 105, 191

Habitats, 13, 18–19, 46–47, 57, 224
Hail, 69
Hand screens, **214**
Harvestmen, 79
Havasu Lake, 183
Headwaters, 81, 155, 162, 165, 171, 177, 184, 224
Hellbenders, **205**
Hellgrammites, **22, 121, 217**
Hemoglobin, 38, 44, 139, 224
Herbivores, 116, 124, 224
Herons, **98–99,** 116, **168,** 183
Hoover Dam, **180–181,** 183
Hosts, 91, 224
Hot Springs National Park (Arkansas), 62, 205
Hudson Highlands, 155
Hudson River, 150–155
Hummingbirds, 61
Hydrogen, 110
Hydropsyche (caddisfly larvae), **24,** 28, 117

Ice, **162–163** (*see also* Glaciers)
Illinois River, 140
Imperial Dam, 183
Industrial wastes, 136–137, 140–141
Insectaria, **214**
Insecticides, 141–143
Insects, 84, 99
  aquatic, **215–219**
  gills of, 44–47, 92
  population of, in rivers, 92, 120–121
  (*See also* Larvae; Nymphs; Pupae; *specific insects*)
Irrigation, 131–134, 170–172, 176, 183
Isonychia (mayfly nymphs), **47**
Itasca, Lake, 162, 164–165

Jack rabbits, 172
Jacks Fork River, 145
Joshua trees, 183
Junipers, 171–172

Katmai National Monument (Alaska), 208
Kentucky Reservoir, 170
Killifish, **105,** 107
Kingfishers, 9–**11**
Kings Canyon National Park (California), 209–210

Lake Mead National Recreation Area (Arizona, Nevada), 180, 183, 208
Lake trout, 160–161
Lakes, 74–75, 84, 97, 134, 155, 159–161, 169
  plankton in, **117**
  in water cycle, 69–71
  (*See also* Reservoirs; *specific lakes*)
Lampreys, **160–161**
Largemouth bass, **39,** 105, 123–125
Larvae, **22,** 30, 138, 140, 224
  breathing of, 42–**47**
  drift and, 117
  fish, 161
  metamorphosis of, 48–50
  on river bottoms, 91–92, 97, **121**
  in swift water, **12–17, 24,** 28, 59, 92
Leafhoppers, 119
Least bitterns, **168**
Levees, 166–167
Lizards, 172
Loons, **74,** 164
Lungs, 38, 224
Lures, fishermen's, **119**

Maggots, 42–43, 138
Mammals, 40, 99
  web-footed, 34, 84, 90
  (*See also specific mammals*)
Mammoth Cave National Park (Kentucky), 73, 203, 208
Mammoth Hot Springs (Yellowstone National Park), **63**
Marsh marigolds, 59–**60**
Marshes, 89, 97, 142, 168
  in water cycle, 70–**71,** 73
Mayflies, 79, **119, 215**
  mating of, **96**–97
  molting of, 51, **54–55**
  nymphs of, 13, 20, **22, 44**–45, **47,** 93, 96, **116, 121,** 138, **215**
Mead, Lake, 180, 183
Meadow mice, 88
Meltwater, 78–79, 176
Merced River, **26**
Mesquite, 172, 176
Metamorphosis, 48–55, 224
  incomplete, 49, 51, 224
Mice, 61, 88, 197
Microorganism, defined, 224
Middle-aged rivers, defined, 81
Midges, larvae of, **24,** 44, **47,** 57, 66, 97, **121,** 139, **219**
  mating of, **97**–98
Migration, of birds, 75, 151, 169–170
  of fish, 31, 105, 151–153, 161, 190–199
Miller's-thumbs (sculpins), 20, 39, **46,** 155
Minerals, 67, 109, 134, 136
Minks, 31, 75, 84, 155
  prey of, 89, 116
Minnows, 102
Mississippi River, 96, 102, **162–167**
  fish in, 105, 190
  floods on, **166–167,** 169
Missouri River, 163–165, 167
Mojave, Lake, 183
Molting, 49, 51–52, **54–55,** 224
Monongahela River, 165
Moose, **64,** 75
Mosquitoes, 38, 138–139, 142, **219**
Moss, 59, 113
Moths, 49
Mount McKinley National Park (Alaska), 203, 208
Mount Rainier National Park (Washington), 76–79, 203, 209
Mountain beavers, **185**
Mountain lions, 172–**173,** 175, 183
Mountain sheep, 64, **177,** 183
Muck (*see* River bottoms)
Mud pots, 63
Muddlers (sculpins), 20, 39, **46,** 155
Mudpuppies, **42**
Mule deer, 172
Mullet, 123–125
Multiple use, defined, 170
Muskrats, 31, 88–90, 94, 109
Mussels, 90–92, 94, 107, 135

Naiads, 51
National Park Service, 145
National Park System, 203–210
National Wild Rivers System, 145–146
Needham, Paul R., 128–131
Nests, bird, 28, **98**
fish, 103, 105, 161
Net-building caddisfly larvae, **121**
Net-winged midge larvae, 24, **47**, 57, **121**
Nets, **20**, 120, 152–153, 191
of insects, 24, 117, **121**
plankton, **212**
Niagara Falls, 14, 160
Nisqually Glacier (Mount Rainier National Park), 76–79
Nisqually River, 78–79
Norris Geyser Basin (Yellowstone National Park), 62
Nostoc (algae), 115, **121**
Nymphs, insect, 13, 20, 96, 138, 140, 224
breathing of, **44–45**, **47**
in food chains, **116**
in hot springs, 66
metamorphosis of, 49, **52–55**
on river bottoms, 92, 95, **121**
in swift water, 13, **22–23**, 28

Oceans, 81, 100
river mouths on, 105–107, 155, 158, 192
in water cycle, 68–69, **71**, 73
(*See also specific oceans*)
Odum, Howard T., 121–124
Ohio River, 149, 163, 165, 167, 169
Oil, waste, **137**
"Old Faithful" (Yellowstone National Park), **62**
Old rivers, defined, 81
Olympic mudminnow, **221**
Olympic National Park (Washington), 209
Olympic Peninsula (Washington), 76
Omnivore, defined, 224
Opossums, 31, **169**
Organic, defined, 224
Otters, 31, 39, 75, 84–88
in food chains, 116–**117**
Owls, 89
Oxygen, 38–45, 97, 140
inadequate supply of, 61, 67, 135–136, 138–139
production of, 109–110
Oysters, **105**, 107
Ozark National Scenic Riverways (Missouri), 145–147, 203, 209

Pacific Ocean, 176, 184, 190, 192
Paddlefish, 105, **165**
Painted turtles, 98, **100**
Parasites, **91**–92, 160–161, 224
Parker Dam, 183
Peccaries, 172, 175
Pecos River, 176
Pelicans, 107
Photosynthesis, 110, 224
Pickerel, **209**
Pickerelweed, 112
Pine trees, 83, 171–172
Pinyon pines, 171–172
Plankton, 117, 224
Plants, 18–19, 24, 36–37
microscopic, 13, 47, 90, 97, 192
(*see also* Algae; Plankton)
water stored in, 69–72
(*See also specific plants*)
Plovers, 107
Pollen, 79

Pollution, 92, 110, 134–145, 197, 224
of Hudson River, 152–153, **155**
purification and, 138–141, 149
Ponds, 75, 97, 103, 117, 131
Pondweeds, **112**
Potomac River, 149, 159
Powell, Lake, 183
Predator, defined, 225
Prickly-pear cactuses, 172
Primary producers, 110, 115–116, 123–125, 127, 225
Production pyramids, 124–**125**, 127, 225
Productivity, defined, 225
measurement of, 119–125, 127–131
Pronghorns, 175
Protozoans, 66, 117
Pupae, 15–17, 48–49, 51, 97, 225
Purification, 138–141, 149, 152–153

Rabbits, 164, 172
Raccoons, 31, **32–33**, 75, 155
Radioactive wastes, 141–143
Rainbow trout, 129
Rainfall, 69, 76, 134, 176, 183
average, 74, 177
floods after, 18–19, **166**, **178–179**
Rapids, 12–17, 47, 92, 192
Rat-tailed maggots, 42–43, 138
Razor clams, 107
Recovery zones, 138–140
Red River, 163, 166
Red-eared turtles, **101**
Redheads, **204**
Refuges, waterfowl, 169–170
Reproduction, 47, 75
of fish, 103, 105, 191–199
of insects, 48–**49**, 51, **94–97**
of mammals, 32, 88–89
of mussels, 91–92
Reservoirs, 134, 167, 170, 172, 180, 183
Riffle bottoms, 120–121
Riffles, 120, 225
Ring-necked ducks, **169**
Ringtails, 175
Rio Conchos, 172
Rio Grande, 145, 149, 170–176
River basins, 163, 165, 167–170, 184, 190
dams of Colorado River, *map* **183**
in United States, *map* **156–157**
River bottoms, 90–94, 95, 97, 112, 139
fish on, 102–103, 105
productivity samples from, 120–121, 123
River mouths, 105–107, 155, 158–159, 166, 176, 192
River otters, 84–88
River patterns, 80–81
Roanoke River, 158
Robber flies, 98
Rock layers, impervious, **70–71**, 72
Rocky Mountain National Park (Colorado), 209
Rocky Mountains, 107, 165–166, 176–177
Rogue River, 149
Rotifers, 117
"Royal coachman" lures, 119

Sacramento River, 177, 184
Sagehen Creek (California), 129–131
St. Anthony, Falls of, 164
St. Lawrence River, 158–160
St. Louis River, 159
Salamanders, 12, **42**, **47**, 61–62
blind, **73–74**
Salmon, 31, 105, 190–199, **221**
Salt water, 105–107, 155 (*see also* Oceans)

San Joaquin River, 177, 184
San Luis Valley (Colorado), 171
Sand bars, formation of, 83
Sandbags, **167**
Sandpipers, **29**
Savannah River, 158
Scarlet lobelia, 72
Scavengers, 116, 225
Scuds, 61
Sculpins, 20, 39, **46**, 155
Seals, 192
Seegrist, Donald, 129
Septic zones, 138–139
Sequoia National Park (California), 209–210
Sewage, 134, 137–138, 153
treatment plants for, **140–141**, 143
Shad, 151–153, **220**
Shallows, 92, 102–103, 109–110, 112
Sharks, 192
Shells, 90–91, **100–101**, 115
Shooting stars (flowers), 72
Shrimps, 107
Sideswimmers, 61
Sierra Nevada Mountains, 76, 129, 184
Silt, 81–83, 92–**93**, 172, 183
as pollutant, 134–135
Silver Springs (Florida), 121–125, 127
Siphons, mussel, 90
Skunks, 61
Sleet, 69
Snails, **37**–38, 40, 61, 102
in food chains, 116, 124–**125**
Snake River, 184
Snapping turtles, **100**, 102
Snow, 69–71, 176–177, 180
Snowpack, 76, 166, 225
Snowshoe hares, 164
Sockeye salmon, 191–192
Softshell turtles, **101**–102
Spawn, defined, 225
Specialization, defined, 225
Species, defined, 225
Sperm, defined, 225
Sphaerotilus (bacteria), 138
Spiracles, 42, 225
Spirogyra (algae), **114–115**
Sponges, fresh-water, 15
Spotted sandpipers, **29**
Spring salamanders, 61–62
Springs, 59–67, 70–73, 225
hot, 62–67
Silver, 121–125, 127
Springtails, 37–38, 79
Stagnant water, 40, 42, 115
Standing crop, 120, 124, 225
Stoneflies, 79, **215**
nymphs of, 13, 20, **22**, 28, 45, 92, **121**, 140, **215**
Streamlined, defined, 225
Sturgeon, 105, 190, **220**
Suckers, 155
Sun, 110, 115, 124, 127
Sunfish, **102–103**, 138
in food chains, **116**, 123–125, 127
Superior National Forest (Minnesota), 74–75
Surber samplers, **120**
Surface film, 36–38
Surface tension, 37, 225
Susquehanna River, 158
Sustained yield, 131, 225
Suwannee River, 149
Swallows, 98, 116
Swamps, 74, 100, 166
Swarms, insect, 96–98
Sycamores, 83

Tadpoles, 23, 42
Tear-of-the-Clouds, Lake, 155

Temperature tolerances, 63, 66, 136
Tennessee River, 163, 165, 177
Tennessee Valley Authority (TVA), 167–170
Terns, 107
Tiger beetle larvae, 30
Toad bugs, 29–**30**
Torrents, 24–28, 192
Tracheae, 38–39, 44, 225–226
Transpiration, 70–72, 226
Travertine terraces, **63**
Tributary, defined, 226
Trout, 12, **47,** 57, 75, 119, 129–**131,** 155
  parasites on, 160–161
  productivity of, 127–131
Tubifex worms, 92, 139
Turtles, 40, 98, **100**–102, 109, 115
  in food chains, 122–125
  prey of, **21,** 102
TVA (Tennessee Valley Authority), 167–170
Two-lined salamanders, **47**

Underground water, 70–73

Vireos, 59

Walleyes, 169
Wastes, industrial, 136–137, 140–141
  radioactive, 141–143
  (*See also* Pollution)
Water beetles, 12, 61, 66, 124–**125**
  breathing of, 42–45
Water boatmen, 43, **217**
Water bugs, creeping, 66
  giant, **216**
Water caterpillars, **44**
Water consumption, 133–134
Water crowfoot, **112**–113
Water cycle, 68–75, 107
Water hyacinths, 36, 112–**113**
Water lilies, 36, 43–44, **110**
Water mites, 61, 66
Water ouzels (dippers), **28**–29
Water pennies, **14**–15, 92, **121, 219**
Water pollution (*see* Pollution)
Water scavenger beetles, **218**
Water scorpions, 44
Water shrews, **35**–37
Water snakes, 12, 98–99
Water striders, 12, **36**–37, **216**
Water table, 70–72, 74, 226
Waterfalls, 24–28, 86, **126,** 190, 192
  (*see also specific waterfalls*)
Waterscopes, **213**
Watersheds, 176–177, 226
Waterthrushes, 29
Weasels, 84
Welland Ship Canal, 158–160
Whirligig beetles, **38, 218**
White water, 83, 226
Willows, 83–84, 119, 172
Winthrop Glacier (Mount Rainier National Park), **78**
Wood ducks, **75**
Woodpeckers, 75
Worms, 61, 73, 92, 139

Yellowstone Falls, **65**
Yellowstone National Park (Idaho, Wyoming, Montana), 30, 62–67, 184, 210
Yellowstone River, **65**
Yosemite Falls, **27**
Yosemite National Park (California), **26–27,** 210
Young rivers, defined, 81
Yuccas, 172

Zion National Park (Utah), 72, 210